NASA STI Program...in Profile

Since its founding, NASA has been dedicated to the advancement of aeronautics and space science. The NASA Scientific and Technical Information (STI) program plays a key part in helping NASA maintain this important role.

The NASA STI Program operates under the auspices of the Agency Chief Information Officer. It collects, organizes, provides for archiving, and disseminates NASA's STI. The NASA STI program provides access to the NASA Aeronautics and Space Database and its public interface, the NASA Technical Reports Server, thus providing one of the largest collections of aeronautical and space science STI in the world. Results are published in both non-NASA channels and by NASA in the NASA STI Report Series, which includes the following report types:

- TECHNICAL PUBLICATION. Reports of completed research or a major significant phase of research that present the results of NASA programs and include extensive data or theoretical analysis. Includes compilations of significant scientific and technical data and information deemed to be of continuing reference value. NASA counterpart of peer-reviewed formal professional papers but has less stringent limitations on manuscript length and extent of graphic presentations.

- TECHNICAL MEMORANDUM. Scientific and technical findings that are preliminary or of specialized interest, e.g., quick release reports, working papers, and bibliographies that contain minimal annotation. Does not contain extensive analysis.

- CONTRACTOR REPORT. Scientific and technical findings by NASA-sponsored contractors and grantees.

- CONFERENCE PUBLICATION. Collected papers from scientific and technical conferences, symposia, seminars, or other meetings sponsored or cosponsored by NASA.

- SPECIAL PUBLICATIONS. Scientific, technical, or historical information from NASA programs, projects, and missions, often concerned with subjects having substantial public interest.

- TECHNICAL TRANSLATION. English-language translations of foreign scientific and technical material pertinent to NASA's mission.

Specialized services also include creating custom thesauri, building customized databases, organizing and publishing research results.

For more information about the NASA STI program, see the following:

- Access the NASA STI program home page at http://www.sti.nasa.gov

- E-mail your question via the Internet to help@sti.nasa.gov

- Fax your question to the NASA STI Help Desk at 443-757-5803

- Telephone the NASA STI Help Desk at 443-757-5802

- Write to:
  NASA Center for Aerospace Information (CASI)
  7115 Standard Drive
  Hanover, MD 21076-1320

NASA/TM—2013-217910

Refractory Materials for Flame Deflector Protection System Corrosion Control:
Similar Industries and/or Launch Facilities Survey

Luz Marina Calle
NASA Kennedy Space Center

Paul E. Hintze
NASA Kennedy Space Center

Christopher R. Parlier
NASA Kennedy Space Center

Brekke E. Coffman
NASA Kennedy Space Center

Jeffrey W. Sampson
NASA Kennedy Space Center

Mark R. Kolody
ASRC Aerospace Corporation

Jerome P. Curran
ASRC Aerospace Corporation

Stephen A. Perusich
ASRC Aerospace Corporation

David Trejo
Texas A&M University

Mary C. Whitten
University of Central Florida

Jason Zidek
Versabar, Inc.

National Aeronautics and
Space Administration
Kennedy Space Center

January 2009

NASA/TM—2013–217910

Available from:

NASA Center for AeroSpace Information
7115 Standard Drive
Hanover, MD 21076-1320

National Technical Information Service
5301 Shawnee Road
Alexandria, VA 22312

NASA/TM—2013–217910

Executive Summary

Refractory concrete is used to protect National Aeronautics and Space Administration (NASA) launch structures from elevated temperatures, ablation, and erosion. The only refractory material qualified for use at Launch Complex 39A (LC 39A) and Launch Complex 39B (LC 39B) is Fondu Fyre WA-1G which is supplied by the Pryor Giggey Co. The material was developed solely for NASA in the 1960s.

Refractory concrete at LC 39A and LC 39B has become susceptible to failure, resulting in large sections of the material breaking away from the base structure. During launch, these sections become high-speed projectiles that jeopardize the safety of KSC personnel, and have the potential to damage ground support equipment and the Space Shuttles.

A review of the current specification and requirements for refractory materials indicates that the test methods and qualification criteria are not well defined. Consequently, the only refractory product qualified for use at the Kennedy Space Center (KSC) may not have the material properties necessary to survive extended exposure to Florida coastal environments and the severe launch conditions exhibited by the Space Shuttle. As a result, better performing refractory materials may be available for use at KSC.

A literature search was conducted to ascertain the different categories of refractory materials that are available for the protection at KSC's launch pads. The classes of materials were categorized as follows:

- Firebrick
- Refractory Concrete
- Silicone and Epoxy Ablatives

Based upon this information, a literature survey was conducted to locate industries that had refractory requirements that were similar to NASA's. Based upon this survey, site visits, and interviews with pertinent industry personnel and refractory vendors were conducted. An analysis of refractory materials at the following locales was then initiated:

- Stennis Space Center
  - A-2 Test Facility
  - B1 and B2 Test Facilities
  - E-2 Test Facility
  - E-3 Test Facility
  - Bldg 3300

- Cape Canaveral Air Force Station
  - Launch Complex 17
  - Launch Complex 34
  - Launch Complex 36
  - Launch Complex 37
  - Launch Complex 40
  - Launch Complex 41

- Vandenberg Air Force Base
  - Launch Complex 6
- Palm Beach County Fire Training Facility

As a result of the site visits and interviews, a series of products for launch applications were found.

Firebrick, while historically used near flame deflectors at NASA launch sites, was not found at any of the locales investigated in this report. Product and labor costs associated with the installation of the materials were cited as the driving factors for its lack of use.

Refractory concrete was used at numerous launch locations. Currently used products include Fondu Fyre WA-1G and Fondag DG. Both versions are gunnable, and as a result, benefit from reduced labor costs associated with the application of the product. Kruzite GR Plus is another refractory concrete that is noteworthy. Kruzite GR Plus is gunnable and provides better adhesion (less rebound) when used in overhead flame duct locations.

The direct impingement areas (of the flame deflectors) were often found to be protected by a ceramic-filled epoxy called Martyte. Martyte was often used to replace and protect refractory concrete that had deteriorated. Furthermore, structural steel (in direct impingement areas) was often protected by the product.

Havaflex is a phenolic ablative that is produced by Ametek Chemical Products. It can be either troweled or sprayed as required, and is used in areas that are subject to direct rocket exhaust.

Various silicone ablative materials were used outside direct blast areas. These coatings included the General Electric GE 3404 ablative, as well as other proprietary formulations from other manufacturers and aerospace companies. These products are often used to protect structural steel, launch pad tubing, and connectors for launch pad instrumentation.

None of the products in this trade study can be considered a panacea for LC 39A and LC 39B. Fondag DG, while inexpensive, was often top-coated with Martyte for repair or additional thermal protection.

Martyte is costly and difficult to apply. Furthermore, incompatibilities between Martyte and the silicone ablatives may be of concern.

Havaflex is a phenolic ablative material that is easy to apply; unfortunately, it is costly and requires frequent replacement.

The silicone ablatives are inexpensive, easy to apply, and perform well outside of direct rocket impingement areas. When used in locations subject to direct rocket exhaust, the performance of the coating is exceeded by refractory concrete and the epoxy alternatives.

This report summarizes the ablative materials that were found at industries with refractory requirements that are similar to NASA's. The refractory products may be considered for use at LC 39A and LC 39B provided the appropriate testing requirements and specifications are met.

Contents

1	INTRODUCTION	1
1.1	Objectives of the Exploration Technology Development Program (ETDP) Refractory Studies	1
1.2	Current Task	2
1.3	Launch Environment	2
2	ASSETS EVALUATED IN THIS ANALYSIS	4
3	DETERIORATION AND REPAIR OF REFRACTORY MATERIAL USED AT LC 39A AND LC 39B	10
4	GENERAL CATEGORIES OF REFRACTORY MATERIALS FOR LAUNCH APPLICATIONS	13
4.1	Firebrick	13
4.2	Refractory Concrete	15
4.3	Epoxy and Silicone Ablatives	16
5	MATERIAL INVESTIGATIONS BY SITE	17
5.1	Stennis Space Center	17
5.1.1	A-2 Test Stand	18
5.1.2	B-1 and B-2 Test Stands	20
5.1.3	E-2 Test Facility	23
5.1.4	E-3 Test Facility	25
5.1.5	Bldg. 3300 Stennis Space Center	27
5.2	Cape Canaveral Air Force Station Launch Facilities	29
5.2.1	Launch Complex 17	29
5.2.1.1	LC 17A	31
5.2.1.2	LC 17B	34
5.2.2	Launch Complex 34	40
5.2.3	Launch Complex 36	42
5.2.4	Launch Complex 37	43
5.2.5	Launch Complex 40	44
5.2.6	Launch Complex 41	45
5.3	Vandenberg Air Force Base: Launch Complex 6	46
5.4	Palm Beach County Fire Training Facility	47
6	CONCLUSIONS	50
APPENDIX A.	PRODUCT DATA SHEETS	55

Figures

Figure 1.	KSC Launch Complex 39B	5
Figure 2.	Space Shuttle Launches by Year and Complex	7
Figure 3.	Cross Section of Flame Deflector at Launch Complex 39A	8
Figure 4.	Openings for Flames From the Main Engine and SRBs	9
Figure 5.	Magnified View of LC 39A Flame Deflector	10
Figure 6.	Evidence That a Section of Concrete Was Dislodged During the Launch of STS-126	11
Figure 7.	Section of Concrete That was Dislodged During the Launch of STS-126	12
Figure 8.	Firebricks at the Base of Launch Complex 34	13
Figure 9.	Firebrick at Launch Complex 34	14
Figure 10.	Damaged Walls at LC 39A	14
Figure 11.	Stennis Space Center Location	18
Figure 12.	Exterior View of the A-2 Test Stand	18
Figure 13.	View of the A-2 Flame Duct and Diffuser	19
Figure 14.	Holes Drilled Into the Surface of the Steel A-2 Test Stand	20
Figure 15.	Rocket Motor Test at Stennis B-1 Complex	21
Figure 16.	Stennis B-2 Flame Deflector	22
Figure 17.	Drain Holes at the Base of the B-2 Flame Deflector	22
Figure 18.	E-2 Cell 1 Test Stand at Stennis Space Center	23
Figure 19.	Stennis Space Center E-2 Cell 2 Vertical Test Flame Duct	24
Figure 20.	Refractory Concrete on Walls and Floor of Stennis E-2 Cell 2 Flame Deflector	24
Figure 21.	E-3 Vertical Test Cell Flame Duct at Stennis Space Center	25
Figure 22.	E-3 Vertical Test Cell Flame Duct at Stennis Space Center – After Firing	26
Figure 23.	Close-up view of Damage to E-3 Cell 2 Test Stand	26
Figure 24.	Fondu Fyre Blocks at the Stennis Space Center E-3 Location	27
Figure 25.	Test Duct at Stennis Space Center	27
Figure 26.	Refractory Concrete Apron at Stennis Space Center	28
Figure 27.	Stennis Plume Deflector Test Rig	29
Figure 28.	Aerial View of LC 17	30
Figure 29.	Major Structures at LC 17	31
Figure 30.	Underside of LC 17A Flame Deflector	32
Figure 31.	Flame Deflector at LC 17A	33
Figure 32.	Fondu Fyre Patched Areas Using Martyte	33
Figure 33.	Side Flame Duct at LC 17B	34
Figure 34.	Concrete Tunnel Adjacent to LC 17B	35
Figure 35.	Damage to Side Flame Tunnel Support Columns	35
Figure 36.	Enclosed Flame Trench at LC 17B	36
Figure 37.	Abrasion to the Interior Surface of Main Flame Trench Walls	36
Figure 38.	Flame Duct at LC 17B	37
Figure 39.	Martyte Patch on Flame Deflector Side Wall	37
Figure 40.	Martyte on Water Deluge System	38

Figure 41.	Martyte-Protected Bolt Heads	38
Figure 42.	LC 17B Launch Deck	39
Figure 43.	LC 17B Exhaust Port on Launch Deck	39
Figure 44.	Ablative Coating on Tubing in Center of LC 17B Launch Deck	40
Figure 45.	LC 34 Flame Deflector and Launch Pad	41
Figure 46.	Flame Deflector at LC 34 on December 11, 2008	41
Figure 47.	Refractory Material at LC 34	42
Figure 48.	Delta IV Heavy at Launch Complex 37	43
Figure 49.	LC 6 at Vandenberg Air Force Base	47
Figure 50.	Palm Beach County Fire Rescue Training Facility	48
Figure 51.	System 203 Tiles at the Palm Beach County Fire Training Center	49
Figure 52.	Interlocking System 203 Tiles	50

Tables

Table 1.	Corrosion Rates of Carbon Steel Calibrating Specimens at Various Locations	3

This page intentionally left blank.

Abbreviations, Acronyms, and Symbols

°F	degree Fahrenheit
AFB	Air Force Base
AFSPC	Air Force Space Command
CCAFS	Cape Canaveral Air Force Station
CRS	Cargo Resupply Services
CxP	Constellation Program
EELV	Evolved Expendable Launch Vehicle
ELV	Expendable Launch Vehicle
ETDP	Exploration Technology Development Program
FOD	foreign objects and debris
FSS	Fixed Service Structure
ft^2	square foot
FUT	Fixed Umbilical Tower
GH_2	gaseous hydrogen
GHe	gaseous helium
GN_2	gaseous nitrogen
GO_2	gaseous oxygen
GOP	Ground Operations Project
gpm	gallon per minute
GSE	ground support equipment
IRBM	intermediate-range ballistic missile
JPL	Jet Propulsion Laboratory
KSC	Kennedy Space Center
lb	pound
LC	Launch Complex
LH_2	liquid hydrogen
LOX	liquid oxygen
MLP	Mobile Launch Platform
MST	Mobile Support Tower
NASA	National Aeronautics and Space Administration
psi	pound per square inch
RSS	Rotating Service Structure
SRB	solid rocket booster
SSC	Stennis Space Center
STS	Space Transportation System
UT	Umbilical Tower

This page intentionally left blank.

1 INTRODUCTION

During the Technology Prioritization Panel held in December 2007, the Constellation Program (CxP) Ground Operations Project (GOP) identified corrosion control technologies as their #2 technology need for initial capability to meet Draft Stretch/Operability requirements for reduced ground processing complexity, streamlined integrated testing, and operations phase affordability.

The *Refractory Materials for Flame Deflector Protection System Corrosion Control* task under the Supportability project will develop refractory technologies that will provide support at Kennedy Space Center (KSC) launch facilities and ground systems through increased operational life cycles.

As a result of the constant deterioration from launch heat/blast effects and aggressive environmental exposure, the refractory materials currently used as a part of the launch pad flame deflectors have become very susceptible to failure, resulting in large pieces of refractory materials breaking away from the steel base structure. These pieces are projected at high speed during launch, and jeopardize the launch complex, vehicle, and safety of the crew.

Replacement refractory systems must be developed to withstand the extremely corrosive environment at the launch pads, and the highly corrosive hydrochloric acid and heat/blast effects that are generated by the solid rocket boosters during a launch. Advanced technologies for the corrosion protection of launch pad flame deflectors are necessary to address these problems and significantly impact ground processing and launch safety.

1.1 Objectives of the Exploration Technology Development Program (ETDP) Refractory Studies

The objective of the ETDP project, *Refractory Materials for Flame Deflector Protection System Corrosion Control* is to develop replacement refractory materials that exhibit long-term resistance to degradation. This degradation results from the extremely corrosive Florida coastal environment and aggressive launch conditions. The highly corrosive solid rocket booster (SRB) exhaust, extreme temperature fluctuations between SRB heat impingement and noise suppression water deluge, and SRB blast vibrations, in combination, have a pronounced detrimental influence on the degradation of refractory materials.

The flame deflector must safely divert flames, exhaust, and small items that are loosened during a launch. In essence, the system must prevent debris from bouncing back and hitting the launch complex and vehicle. Performance in this regard is dependent upon integrity of the refractory materials used on the flame deflectors.

The development process for the ETDP *Refractory Materials for Flame Deflector Protection System Corrosion Control* project has four primary elements.

- Capability to develop a refractory protection system for the launch pad flame deflectors.
    - Long term corrosion protection
    - Mitigate the safety risk caused by frequent failures and unacceptable performance (cracking and spalling during launch) of the current refractory concrete materials.
    - Mitigate the risk of frequent, expensive, and extensive repairs that with the current material, also provide unacceptable performance.
- Capability to develop advanced refractory materials.
    - That provides acceptable performance and maintain their integrity during/after exposure to the launch environment (high temperature exhaust impingement, blast loading, water deluge delta temperature, and acoustic loading) without cracking or spalling.
    - That can resist the degradation of thermal-protection characteristics caused by seacoast exposure.
- Capability to develop material requirements, system specifications, and qualification standards for the refractory material protection system.
- Capability to incorporate the refractory material formulation onto the flame deflector base structure, and evaluate the in-situ performance in an integrated demonstration on a scaled, simulated flame deflector.

Knowledge gained from the development of refractory material systems for flame deflectors will be leveraged to evaluate materials and systems for the replacement of refractory firebricks along the flame trench vertical walls.

1.2 Current Task

The current task consists of a report on a trade study that investigates refractory materials used in similar industries that would provide a direct benefit to KSC launch pad flame deflectors. Understandably, the environment that refractory materials are subject to (under launch) are extreme, and consequently, this investigation is oriented toward materials that are subject to high temperatures, extreme temperature fluctuations, significant erosion and ablation, water infiltration and acidity from the solid rocket boosters.

Site and literature investigations, as well as interviews with key refractory and launch personnel were conducted to ascertain the refractory and ablative materials that are used in these extreme environments. This report summarizes the refractory and ablative materials that are used at other launch locations and facilities that are similar to the environments NASA's launch complexes.

1.3 Launch Environment

The launch facilities at KSC are approximately 1000 feet from the Atlantic Ocean. The seacoast marine location is extremely corrosive to structural steel. In fact, the beachside location at KSC is documented as one of the most corrosive environments in the world. Table 1 shows the corrosion rates for the KSC Beachside Exposure Corrosion Test Site. The corrosion rates in the table clearly show the aggressiveness of the KSC locale, in relation to the others that are listed.

Table 1. Corrosion Rates of Carbon Steel Calibrating Specimens at Various Locations[1]

Location	Type of Environment	μm/yr	mils/yr
Esquimalt, Vancouver Island, BC, Canada	Rural marine	13	0.5
Pittsburgh, PA	Industrial	30	1.2
Cleveland, OH	Industrial	38	1.5
Limon Bay, Panama	Tropical marine	61	2.4
East Chicago, IL	Industrial	84	3.3
Brazos River, TX	Industrial marine	94	3.7
Daytona Beach, FL	Marine	295	11.6
Pont Reyes, CA	Marine	500	19.7
Kure Beach, NC (24 m from ocean)	Marine	533	21
Galeta Point Beach, Panama	Marine	686	27
Kennedy Space Center, FL (Beach)	**Marine**	**1070**	**42**

KSC launch facilities and ground support equipment (GSE) are exposed to extremely corrosive marine conditions. As if those conditions were not bad enough, in 1981, the Space Shuttle introduced a more aggressive environment to the launch pads at KSC. Exhaust from the SRBs resulted in the deposition of small alumina particles with hydrochloric acid adsorbed onto their surface. It is estimated that 70 tons of hydrochloric acid are generated during a Space Shuttle launch. The impingement of this acidic exhaust results in the failure of refractory materials, despite the fact that a pressure wash-down is performed immediately after launch.

In response to the SRB exhaust problem, studies were conducted at KSC to increase the chemical resistance of protective coatings and materials in response to this more aggressive propulsion system.[2,3,4,5,6,7,8]

---

[1] Coburn, S., "Atmospheric Corrosion," in American Society for Metals, Metals Handbook, Properties and Selection, Carbon Steels, Metals Park, Ohio, 9th ed., Vol. 1, p.720, 1978.

[2] Ruggieri, D. and Rowe, Anne, "Evaluation of Carbon Steel, Aluminum Alloy, and Stainless Steel Protective Coating Systems After 18 Months of Seacoast Exposure," NASA Technical Memorandum 103503, May 1984.

[3] MacDowell, L.G., "Evaluation of Protective Coating Systems for Carbon Steel Exposed to Simulated SRB Effluent after 18 months of Seacoast Exposure," NASA Report No. MTB-268-86B, February 1988.

[4] MacDowell, L.G., "Volatile Organic Content (VOC) Compliant Coating Systems for Carbon Steel Exposed to the STS Launch Environment – Application, Laboratory and 18 Month Exposure Results," NASA Report No. FAM-93-2004, February 23, 1993.

[5] MacDowell, L.G., "Testing VOC-Compliant Coating Systems at Kennedy Space Center, Materials Performance," 32, p. 26–33, 1993.

[6] Calle, L.M., and MacDowell, L.G., "Improved Accelerated corrosion Testing of Zinc-Rich Primers," NASA Tech Briefs, 24, p. 78, 2000.

[7] Calle, L.M., and MacDowell, L.G., "Evaluation of Inorganic Zinc-Rich Primers Using Electrochemical Impedance Spectroscopy (EIS) in Combination with Atmospheric Exposure," in proceedings of NACE International Conference on Corrosion in Natural and Industrial Environments: Problems and Solutions, Grado (Gorizia), Italy May 23–25, 1995.

2 ASSETS EVALUATED IN THIS ANALYSIS

KSC launch operations consist of seven major structures that include two launch complexes, three Mobile Launch Platforms (MLPs), and two Mobile Crawlers. Collectively, these assets consist of 5,443,696 ft^2 of structural steel.[9] Refractory materials were developed and integrated as components to protect these assets from the repeated rocket blast and high temperatures that are exhibited during launch.

Refractory degradation is not limited to the launch complexes at KSC. Consequently, the development of new refractory products will prove beneficial, not only for assets under investigation as a part of the refractory project, but also for assets at other centers. In a similar fashion, the beneficial attributes of new refractory systems can be extrapolated to other government entities and to private industry.

Two complexes support mobile launch operations and are available for the Space Shuttle. Launch Complex (LC) 39A and LC 39B are *sisters* of each other, and share similar characteristics. The launch complex can be tentatively delineated by four sections (Figure 1). The flame deflector, which is the subject of this report, is a subcomponent under pad structures. The four sections include the:[10]

a. Fixed Service Structure (FSS)

b. Rotating Service Structure (RSS)

c. Perimeter

d. Pad Structures

---

[8] Calle, L.M., and MacDowell, L.G., "Evaluation of Inorganic Zinc-Rich Primers Using Electrochemical Impedance Spectroscopy (EIS) in Combination with Atmospheric Exposure," NASA Report No. 94-2082, John F. Kennedy Space Center, Florida, April 17, 1995.
[9] NASA NE-M9, "KSC Corrosion Control Overview for Stennis Space Center," March 13, 2008.
[10] Launch Complex 39, Pads A and B, www.nasa.gov/pdf/168440main_LC39-06.pdf. Last accessed on April 7, 2008.

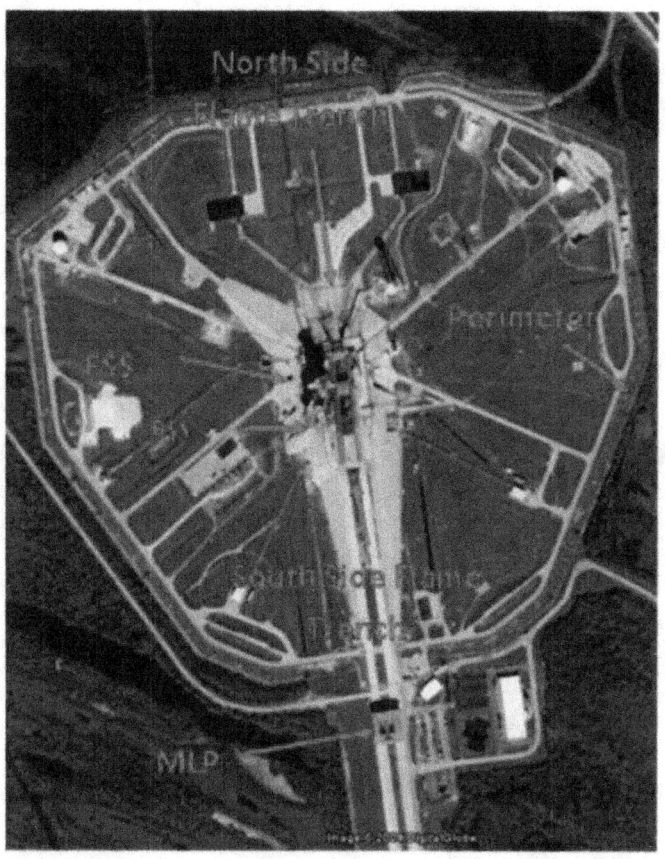

Figure 1. KSC Launch Complex 39B

The FSS and MLPs were originally designed for a 20-year lifespan with the Apollo era vehicles that used liquid propellants. Even after 40 years of service, the structures remain in use with the Space Shuttle and its deleterious solid rocket boosters.[11]

The FSS has three swing arms that provide services and access to the Space Shuttle prior to launch. The Orbiter Access Arm is a component of the FSS that allows personnel to enter the Shuttle crew compartment, and serves as an escape route for the astronauts prior to launch.

The External Tank Hydrogen Vent Umbilical and Intertank Access Arm on the FSS support tanking, and the External Tank Gaseous Oxygen Vent Arm allows for the transfer of heated gaseous nitrogen to warm the liquid oxygen vent system on top of the external tank. This prevents the buildup of ice, which could have damaging consequences to the Shuttle during launch.

The FSS also contains the Emergency Egress System, which includes seven baskets for the emergency evacuation of the launch complex by the astronauts.

---

[11] Personal communication with Harry Moore, United Space Alliance. April 10, 2008.

The RSS provides a means to install and service Space Shuttle payloads while at the pad. Furthermore, this structure supports servicing operations on the Space Shuttle that cannot be performed from the FSS.

The perimeter of the launch pad is the area inside the fence but beyond the pad surface. It includes the liquid oxygen (LOX) and liquid hydrogen (LH_2) storage tanks, as well as pipes, tanks and small buildings. The LOX and LH_2 tanks contain and deliver the fuels to the Space Shuttle's external tank prior to launch.

Launch pad surface structures typically include the MLP holddown posts, the MLP access towers, the hydraulic elevator and the flame deflector.

Figure 2 shows the Space Shuttle launches by pad for each year since the inception of the Shuttle program in 1981. The data proved useful since it helped delineate years when launch pads were not in use, and others when they were used much more extensively. This data can be correlated to times in which more frequent repair to the refractory systems was required. Furthermore, the frequency of launch can be compared to other launch sites.

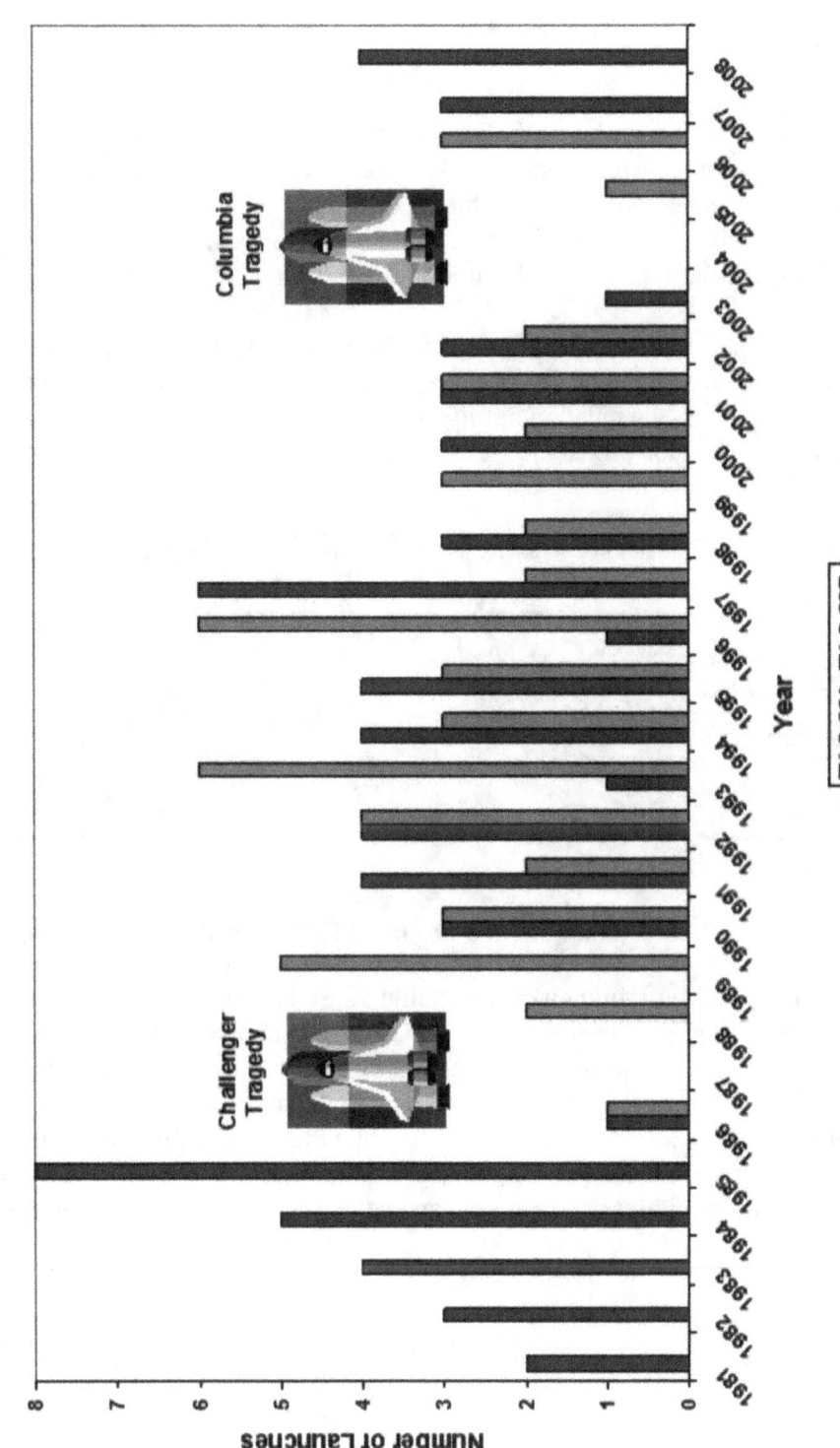

Figure 2. Space Shuttle Launches by Year and Complex

The first Shuttle launch on April 12, 1981, resulted in a few unexpected problems. For example, an anticipated pulse pressure of 0.5 pound per square inch (psi) was actually 2.0 psi when the two SRBs ignited. The redesign of the flame deflector was one of the potential resolutions to the problem. However, a new water suppression system was implemented to address the pressure problem.[12]

Figure 3 shows a schematic of the flame deflector at LC 39. LC 39A and LC 39B were originally designed to support the Apollo program. With the advent of the Shuttle program, the Saturn-era flame deflectors were replaced. The orbiter side of the new flame deflectors are 38 ft high, 72 ft long, and 57 ft wide. The SRB side of the flame deflector is 42 ft high, 42 ft long, and 57 ft wide. The total mass of the asset is over 1 million pounds.[13]

The flames from the main engines and the SRBs are channeled down opposite sides of the flame deflector. The deflector is made of steel on a structural steel I-beam framework. To protect the structure from serious degradation during launch, the faces of the flame deflector are lined with refractory concrete. This product is known as Fondu Fyre WA-1G supplied by the Pryor Giggey Co.

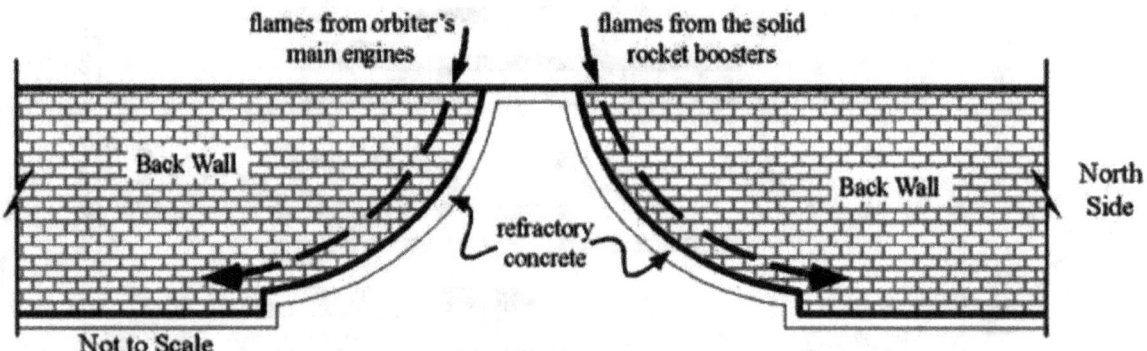

Figure 3. **Cross Section of Flame Deflector at Launch Complex 39A**

Figure 4 shows the configuration of the Shuttle viewed upward from the floor of the flame trench. The openings for the Space Shuttle exhaust and the flame deflector, which is used to divert the rocket plume from the SRBs, are labeled. The other side of the flame deflector, which is not visible in the picture, is for exhaust from the main engines. The SRBs burn at approximately 3000 degrees Fahrenheit (°F), while the exhaust from the main engines burn at a lower temperature. Consequently, the higher temperatures of the SRB exhaust lead to more severe exposure conditions and result in damage that is more significant to the deflector.

---

[12] Kolcum, Edward H., "NASA Studying Pressure Problem in Space Shuttle," *Aviation Week & Space Technology*, July 6, 1981, p.21.
[13] http://science.ksc.nasa.gov/facilities/lc39a.html. Last accessed on December 17, 2008.

Figure 4.   Openings for Flames From the Main Engine and SRBs

Figure 5 shows a magnified view of the flame deflector underneath the SRBs. The bottom of the deflector shows the structural steel, which is protected with Fondu Fyre at a depth of approximately 6 inches.

The ability of the flame deflector to safely meet the requirements of diverting the flame, exhaust, and small items that are dislodged during launch is dependent on the integrity and performance of the refractory materials. Consequently, the use of refractory products that have superior material characteristics (under launch conditions) is required to protect the flame deflector, Space Shuttle, GSE, and launch personnel.

Figure 5. Magnified View of LC 39A Flame Deflector

3 DETERIORATION AND REPAIR OF REFRACTORY MATERIAL USED AT LC 39A AND LC 39B

The launch complexes at KSC are critical support facilities that are required for the successful launch of space-based vehicles. Most of these facilities are over 25 years old, and consequently, are experiencing deterioration. As a result of the constant deterioration from launch heat/blast effects and environmental exposure, the refractory materials used at LC 39A and LC 39B have become susceptible to failure, resulting in large sections of refractory material breaking away from the base structure and creating high-speed projectiles during launch. These projectiles jeopardize the safety of the launch complex, crew, and vehicle. Postlaunch inspections have revealed that the number and frequency of repairs, as well as the area and size of the damage, is increasing with the number of launches.

It is assumed that the composition of Fondu Fyre may have changed. This conclusion is based upon the change in color of the product.

Refractory concrete protects launch structures from elevated temperatures. These materials utilize hydraulic cement as a binding agent. The refractory material used at LC 39A and LC 39B is Fondu Fyre WA-1G supplied by the Pryor Giggey Co. The product was developed in the 1960s solely for the National Aeronautics and Space Administration (NASA) and is the only product qualified for use at LC 39A and LC 39B.

During the selection of the material, the acidic exhaust of the SRBs and the temperature fluctuations that result from the sound suppressing water deluge were not considered. An analysis of test data taken from 1981 to 1993 indicates that this refractory material does not meet the requirements of KSC-SPEC-P-0012, *Specification for Refractory Concrete* (1979). In fact, testing during this period indicated that none of the submitted refractory products could meet the required specifications. As such, the only qualified material for use at KSC does not meet the required specifications. Review of the current specification and testing requirements indicate that the test methods and qualification criteria are not well defined. Other, possibly better-performing materials have not been qualified because of the poorly defined specification tests and requirements.[14]

Failure to meet these standards increases the likelihood that the product will prematurely degrade. This has resulted in cracking of the product, corrosion of the metallic anchoring structure (grid steel and Nelson Ties), spalling, and liberation of the refractory concrete during launch.

Figure 6 shows a section of Fondu Fyre WA-1G refractory concrete that spalled on the main engine side of the flame deflector LC 39A during the launch of STS-126 (November 14, 2008). This section dislodged during launch and was subsequently hurtled downrange.

Figure 6.   Evidence That a Section of Concrete Was Dislodged During the Launch of STS-126

The anomaly may have resulted from the seepage of water through the cracked refractory concrete. This resulted in corrosion to the grid steel, which reduced the adhesion between the refractory and base material. During launch, the water under the concrete section may have turned to steam, further lifting the section from the surface. The thrust from the Space Shuttle main engines hurled the object downrange.

---

[14] Calle, L.M., Trejo, D., and Rutkowsky, J., "Evaluation of Alternative Refractory Materials for the Main Flame Deflectors at KSC Launch Complexes," NASA TM-2006-214197, March 2006.

After the launch of STS-126, the section of refractory concrete was located halfway down the length of the flame trench. A picture of the dislocated concrete section is shown in Figure 7.

Figure 7. **Section of Concrete That was Dislodged During the Launch of STS-126**

Advanced technologies for material protection are required to address problems that impact ground processing and launch safety. The ability of the flame deflector to meet the requirements of diverting the flames, exhaust, and loosened items during launch is critical. The flame deflector must prevent foreign objects and debris (FOD) from deflecting off surfaces and possibly hitting the launch complex and vehicle.

NASA has plans to update LC 39A and LC 39B for the future Ares vehicles. Modifications to the facilities will require reconditioning of the currently used flame deflectors. Ideally, refurbishment will be performed with refractory materials that will extend the useful life of the structure for an additional 40 years of service.[15]

Plans to mount the Ares rocket over the left SRB hole are being considered, and there are concerns related to thrust pressure blowing through the two unused holes (i.e., the right SRB hole and the hole for the main engine). Flame trench pressurization data (from the first Shuttle mission) prior to the implementation of the water suppression system, as compared to data after the water deluge was implemented, predicts that the current flame deflector configuration would not have a pressure increase greater than 2 psi.[16]

---

[15] Coppinger, R., "Ares I-1 pad 39B to be modified," *Flight International*, February 6, 2007.
[16] Mecham, M., "Screen Savior," *Aviation Week & Space Technology*, Vol. 169, No. 7, p. 87, August 18, 2008.

4 GENERAL CATEGORIES OF REFRACTORY MATERIALS FOR LAUNCH APPLICATIONS

4.1 Firebrick

Firebrick is a refractory product that is kiln baked prior to placement. Firebrick contains up to 44% alumina, is dense, and melts at high temperatures.[17] A thorough discussion regarding the material characteristics of firebrick is beyond the scope of this report, though it should be emphasized that the material has been used throughout NASA launch history.

Historical remnants of firebricks can be found at the abandoned LC 34 at KSC (Figure 8). As shown in the figure, the firebricks were produced by the A.P. Green Company. The date of manufacture is the mid to late 1950s.

Figure 8.   **Firebricks at the Base of Launch Complex 34**

While the design of LC 34 necessitated the use of a flame deflector, refractory firebrick was used in non-impingement, elevated temperature areas at the base of the launch structure. The rails shown in Figure 9 were used to transport the flame duct to and from the facility. During launch, the blast was directed to the fore and aft orientation in Figure 9. Consequently, refractory firebrick was designated for these locations. In contrast, normal construction grade concrete was used at the sides.

---

[17] Lee, C.C., and Dar Lin, S., *Handbook of Environmental Engineering Calculations,* McGraw-Hill, 2000. Online version available at
http://knovel.com/web/portal/browse/display?_EXT_KNOVEL_DISPLAY_bookid=621&VerticalID=0

Figure 9.   Firebrick at Launch Complex 34

The use of refractory firebrick for launch pad applications has continued through the Shuttle era. As shown in Figure 10, firebrick was designed into the floors and walls of the flame trench at the Shuttle launch pads.

On May 31, 2008, serious damage to the walls of the flame trench occurred during the launch of the Space Shuttle Discovery (STS-124). Blast from the SRBs resulted in the expulsion of over 500 firebricks from the flame trench walls (Figure 10). The interlocking firebricks measured 6 by 3 by 13.5 inches and were attached to the 3-foot-thick concrete substructure with epoxy and metal clips. The clips, anchored in concrete, are horizontally attached to every other brick, and vertically at every sixth row.

Figure 10.   Damaged Walls at LC 39A

While precast firebricks have improved material characteristics, the placement of the individual pieces into a fully operational design can be complex. Furthermore, structural repair and replacement is cost-prohibitive given the labor costs associated with skilled bricklayers who are required for the appropriate installation of the product. Consequently, while the refractory firebrick provides improved material properties, the cost of the product and labor for installation are much higher than that required for the installation of traditional refractory concrete.[18]

4.2 Refractory Concrete

The selection of refractory concrete materials for flame deflectors requires products that are resistant to thermal shock, provide strength at high temperature, have limited changes in dimension at very high temperature, are spall resistant, resists crack propagation, are resistant to acoustic shock, have insulating properties (steel reinforcement will expand and crack refractory material), and can be cured by normal procedures without cracking.[19] Furthermore, manufactures formulate refractory products so that they can be applied by the Gunite process. Consequently, the product can be placed with little effort as compared to that necessary with the placement of firebrick.

The use of calcium aluminate aggregates, in conjunction with the calcium aluminate cement, produces a hardened concrete that is resistant to the deleterious effects of high temperature environments. When the calcium aluminate aggregate is employed, the shotcrete system is a high-strength product that provides outstanding resistance to heat and thermal shock (to 2000 °F), and very good abrasion resistance.

Calcium aluminate concrete products are used as linings in fire training structures and have been employed in sulphur pit applications in petrochemical industries. Direct contact with molten sulfur produces an extremely corrosive environment at moderately high temperatures.[20] This environment would severely damage structures erected using construction grade Portland Cement containing concrete.

Calcium aluminate cement is used as a binder for refractory concretes that are used in launch environments. Often, the product is used in direct flame impingement areas that include the flame deflectors and exhaust tunnels. Both areas are subject to very high temperatures and abrasion from solid airborne particulates. The calcium aluminate products are extensively used at KSC, Stennis, and Vandenberg Air force Base. The formulation of the concrete mixes can differ between manufacturers.

Typical refractory concretes include Fondu Fyre, which is used for the Space Shuttle at LC 39A and LC 39B, as well as Fondag which is used in other areas of KSC and Vandenberg. Both products use calcium aluminate cement as a binder. Crushed firebrick is added to Fondu Fyre to

---

[18] Telephone interview with Doug Goddard, Atlantic Firebrick, January 6, 2009.
[19] Lays, E.J., and Darrow, E.A., "Effects of Exhausts from Aluminized Solid Propellants on Launch Facilities," *J. Spacecraft*, Vol. 4, No. 7, 934–940, July 1967.
[20] Fitzgerald, M.W., Talley, J., and Alt, C.W., "Calcium Aluminate Technology and Its Application in Refractory Concrete," Shotcrete, Summer 2002.

produce the concrete mix, while Fondag uses a calcium aluminate aggregate.[21] Both products were prevalent at the launch sites visited, and can be applied using the Gunite process.

Refractory materials for launch applications have unique material requirement such as thermal shock, abrasion, and erosion resistance. Consequently, various materials have been tested under simulated launch conditions. Some of these materials include Fondu Fyre WA-1, Fondu Fyre XB-1, Fondag DG, Harbison-Walker Harcast ES, and Harbison-Walker 13-65 Fused Silica Castable Mix.[22,23]

As discussed in later sections of this report, the gunnable formulation of Fondu Fyre (Fondu Fyre WA-1G) and the Kerneos refractory (Fondag DG) were predominantly seen at the launch sites inspected as a part of this report.

4.3 Epoxy and Silicone Ablatives

Protective coatings have been formulated to protect metal structures that are not subject to direct rocket motor blast. Development criteria necessitated that they were easy to apply, adhered well to the underlying substrate, had high resistance to thermal and acoustic shock, and provided erosion and ablation limiting characteristics. Furthermore, the coatings had to be compatible with exhaust residue from the motors, as well as the liquid propellants that are used for launch and vehicle operations. These include LOX and the hypergol propellants.

The two classes of materials that have been considered for these applications include the epoxies and silicones.

Off-the-shelf thermal protective coatings of this type have been evaluated for peripheral launch pad applications. They include, but are not limited to, Dynatherm E-300, Dow Corning Q90-006, Dow Corning Q20-103, Dow Corning Q30-121, Dow Corning Q93-019; Fuller Fulblate 878 Types I and II, Fuller 190J-4, Korblate 11-190:L, General Electric RTV 511 and RTV 757 (foamed); and Raytheon RPR 2138, RPR 2141, RPR 2156, and General Electric CPC-1050 and SCM 3404.[24,25]

While most products in this category were originally designed for use outside direct blast areas, others have been developed for the more aggressive exposure. These products include Martyte and Havaflex TA 117.

---

[21] Telephone interview with Greg Wallace, Kerneos Aluminate Technologies, January 6, 2009.
[22] Lays, E.J., and Darrow, E.A., "Effects of Exhausts from Aluminized Solid Propellants on Launch Facilities," *J. Spacecraft*, Vol. 4, No. 7, 934–940, July 1967.
[23] Douglas, F.D., Dawson, M.C., and Orlin, P.A., "ASRM Subscale Plume Deflector Testing," AIAA 92-3919, AIAA 17th Aerospace Ground Testing Conference, Nashville, Tennessee, July 6–8, 1992.
[24] Lays, E.J., and Darrow, E.A., "Effects of Exhausts from Aluminized Solid Propellants on Launch Facilities," *J. Spacecraft*, Vol. 4, No. 7, 934–940, July 1967.
[25] Sprayable Silicone Ablative Coating, GE Product Code CPC-1050, Post Launch Evaluation on Launch Complex 40, Cape Canaveral Air Force Station, NASA Failure Analysis and Materials Evaluation Branch; Report 92-2150, November 23, 1992.

Martyte has been evaluated as a protective coating for metal surfaces.[26] Martyte is a ceramic filled, amine-cured epoxy compound that was originally developed by Martin Marietta. Site visits revealed that the material was often used on top of construction grade concrete, refractory concrete, and structural steel. Use of this material is described later in this document.

Havaflex TA 117 is a two-component modified phenolic ablative coating and adhesive that was originally developed for the U.S. Navy. The product can be troweled or sprayed, and was designed to protect decks, bulkheads, and shipboard launching systems from the extremely high temperatures (up to 5000 °F.) and high gas velocities that are present during launch operations.[27]

5 MATERIAL INVESTIGATIONS BY SITE

A literature survey and site visits were conducted to determine which refractory materials were used in industries similar to NASA's launch environments.

This section summarizes the results from these investigations.

5.1 Stennis Space Center

Stennis Space Center (SSC) is located in Hancock County, Mississippi, at the Mississippi-Louisiana border (Figure 11). Stennis is NASA's largest rocket engine test facility.

Construction of the 13,500-acre complex began in October 1961. The test area is surrounded by a 125,000-acre acoustical buffer zone. The facility's large concrete and metal test stands were originally used to test-fire the first and second stages of the Saturn V rockets and are now used to flight certify the Space Shuttle main engines.

The site was originally selected by the U.S. government because it was located in a thinly populated area that had barge access. Furthermore, the site is advantageously located between the Michoud Assembly Facility and the launch facility in Cape Canaveral in Florida.

---

[26] Lays, E.J., and Darrow, E.A., "Effects of Exhausts from Aluminized Solid Propellants on Launch Facilities," *J. Spacecraft*, Vol. 4, No. 7, 934–940, July 1967.
[27] Havaflex T.A.-117, A Trowelable Ablative Material. http://new.ametek.com/content-manager/files/HAV//Havaflex1.pdf. Last accessed on January 9, 2008.

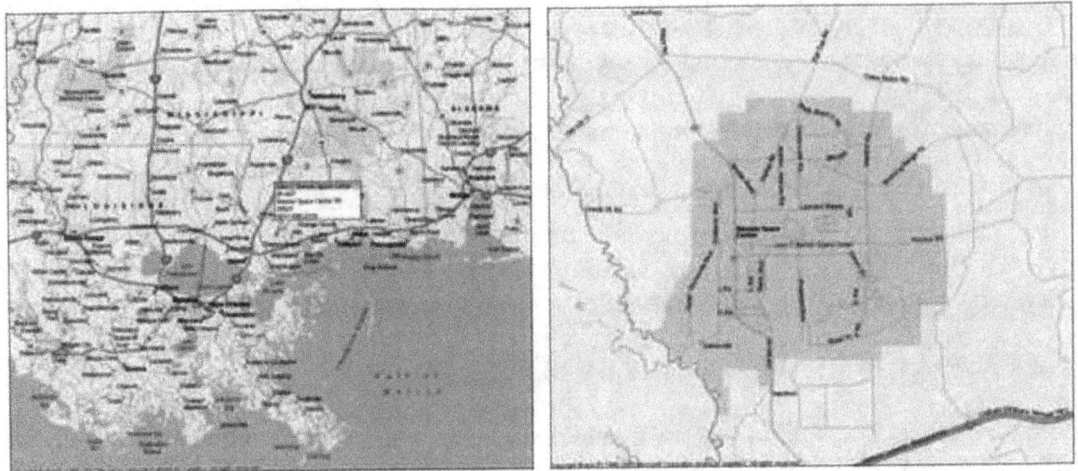

Figure 11.   Stennis Space Center Location

5.1.1   A-2 Test Stand

The A-2 Test Stand is a single-position, vertical firing fixture that can accommodate test articles up to 33 ft in diameter. An exterior view of the test complex is shown in Figure 12.

Figure 12.   Exterior View of the A-2 Test Stand

The test facility is designed to use LH_2 and LOX propellants and can accommodate support fluids, which include gaseous helium (GHe), gaseous hydrogen (GH_2), and gaseous nitrogen (GN_2). It is equipped with an altitude diffuser, which is utilized to simulate altitude conditions during engine testing. The maximum dynamic load that the structure is capable of testing is 1.1 million foot pounds.[28]

Figure 13 shows the relative size of the flame deflector in relation to members of the refractory site review team.

Figure 13. View of the A-2 Flame Duct and Diffuser

The A-2 Test Stand is fabricated from steel and uses water for cooling. Prior to a test, the interior cavity of the flame deflector is filled with water. Thirty minutes before the test, the pipes and manifolds are filled at a rate of 3000 gallons per minute (gpm). Six minutes prior to the test, large pumps increase the flow rate to 160,000 gpm. This produces approximately 225 psi of water pressure.

During the test, water exits from the holes in the surface of the deflector to reduce temperature effects that are induced by the rocket exhaust (Figure 14). Each hole is 5/32 inches in diameter, flows 7 gpm, and is drilled so that there are 14 holes per square foot, in the lower "hot" area. Fewer holes are present on the sides and away from the plume impact.[29] The perforations are also advantageous since they may reduce stress that can warp the metal deflector.[30]

---

[28] A-2 Test Stand. http://sscfreedom.ssc.nasa.gov/esd/ESDTestFacilitiesA2.asp. Last accessed on November 7, 2008.
[29] E-mail correspondence with Nickey Raines, Stennis Space Center, January 14, 2009.
[30] Personal interview with Dr. Charles E. Semler, Semler Material Services, December 1, 2008.

Figure 14.   Holes Drilled Into the Surface of the Steel A-2 Test Stand

In general, some kind of thermal protection is required to reduce thermal stress and buckling of the flame deflector.[31] Early work on deflector design for SRBs, indicated that metallic structures were only feasible if the steel structure remained at least 1.5 nozzle exit diameters laterally from the exhaust plume and provided the vehicle exits from the launch pad surface by at least 50 nozzle diameters in less than 10 seconds.[32]

5.1.2   B-1 and B-2 Test Stands

Each B-1 and B-2 Test Complex consists of a dual-position, vertical, static-firing test stand. The B Complex is 295 ft tall and is equipped with a 200-ton main derrick-lifting crane. The test stand was designed to use LH_2 and LOX propellants, and can accommodate various support fluids that include GHe, GH_2, and GN_2. The maximum dynamic load that each structure is capable of testing is 11 million foot pounds.[33]

The refractory material team from KSC intended to inspect one of the test stands. Unfortunately, a direct inspection at that time was impossible since the area was cleared for a test. The rocket test is shown in Figure 15.

---

[31] Design Handbook for Protection of Launch Complexes from Solid Rocket Propellant; Report No. Martin-CR-66-11; Martin Marietta; March 1966.
[32] A Brief Study of Flame Deflection, Report 1323, Aerojet General, August 1957.
[33] B-1 Test Stand. http://sscfreedom.ssc.nasa.gov/esd/ESDTestFacilitiesB1.asp. Last accessed on November 7, 2008.

Figure 15.   Rocket Motor Test at Stennis B-1 Complex

As a part of a prior project, NASA Corrosion Laboratory personnel photo-documented the B-2 Test Stand flame deflector (Figure 16).

Similar to the A-2 Test Stand discussed in 5.1.1, the B-2 test stand is a steel structure that is much larger in scale. Figure 16 is used to give a comparative size of the B-2 flame duct in relation to others that are discussed in this report. To cool the flame deflector, the deluge system supplies 333,850 gpm of high-pressure industrial water for an extended period at a pressure of 225 psi.[34]

---

[34] Fisher, M.F., and King, R.F., "Low-Cost Propulsion Technology Testing at the Stennis Space Center-Propulsion Test Article and the Horizontal Test Facility," AIAA-98-3367, 34th AIAA/ASME/SAE/ASEE Joint Propulsion Conference and Exhibit, Cleveland, Ohio, July 13–15, 1998.

Figure 16. Stennis B-2 Flame Deflector

The design of the B-2 Test stand was similar to the A-2 test stand previously discussed, as both had holes perforated in the surface of the deflector. The utility of these holes was discussed in Section 5.1.1. Particular mention is made of corrosion in areas adjacent to the B-2 Test stand drain holes (Figure 17).

Corrosion near the drain holes is a particularly important observation with regard to the design of flame deflectors. Flame deflectors that are used for corrosive SRBs in a coastal environment (such as KSC) may suffer from much greater corrosion problems than those seen (Figure 17) at Stennis Space Center, due to the acidic conditions and proximity to the seacoast.

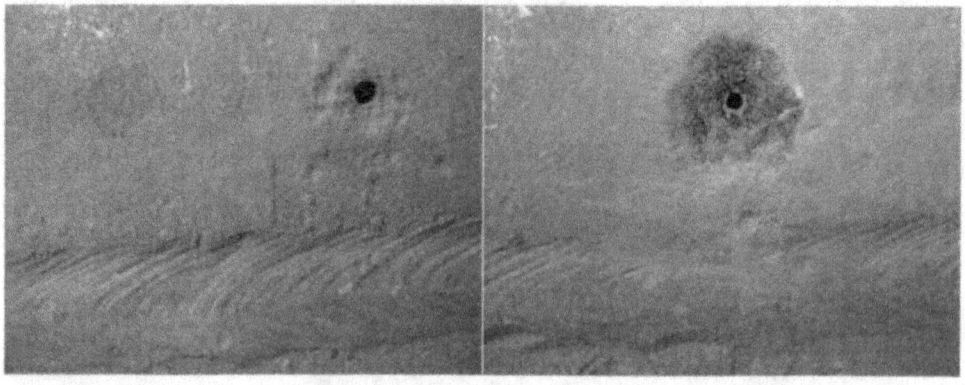

Figure 17. Drain Holes at the Base of the B-2 Flame Deflector

5.1.3 E-2 Test Facility

The E-2 Test Facility was constructed to support materials development by subjecting test articles (including refractory concrete) to extreme temperature conditions and fluctuations. This facility has support capabilities, which include hot gas, cryogenic fluids, gas impingement, inert gases, industrial gases, specialized gases, hydraulics, and water.

The E-2 facility is a multi-cell complex that is capable of testing intermediate size engines in both the vertical and horizontal configurations.

E-2 Cell 1 (Figure 18) is the horizontal test cell, and utilizes propellants such as LOX, LH_2, and RP-1. The horizontal test cell is capable of testing motors with thrust loads up to 120,000 foot pounds.[35] Cell 2 is the vertical test cell, and utilizes LOX and RP-1 propellants. This test cell is capable of testing motors with up to 100,000 foot pounds of thrust.[36]

Figure 18.   **E-2 Cell 1 Test Stand at Stennis Space Center**

Figure 19 shows an overview of the Cell 2 flame duct. As shown in the photo, the flame deflector was built from steel over an I-beam steel structure. The facility has a 4000 gpm water deluge system that is used to protect the flame duct from plume radiant heating during testing.[37]

---

[35] Test Facilities Capability Handbook, Stennis Space Center, NP-2001-11-00021-SSC, November 2001.
[36] E-2 Test Facility. http://sscfreedom.ssc.nasa.gov/esd/ESDTestFacilitiesE2.asp. Last accessed on November 7, 2008.
[37] Test Facilities Capability Handbook; Stennis Space Center; NP-2001-11-00021-SSC; November 2001.

Figure 19.   Stennis Space Center E-2 Cell 2 Vertical Test Flame Duct

The depth of the refractory concrete used to protect the underlying steel varied. The average depth of the refractory layer was approximately two inches at the center. In contrast, the depth of the refractory material at the walls was approximated at only 1 inch in depth.

Figure 20 shows the floor and wall of the Stennis E-2 Cell 2 flame deflector. Site personnel were unable to determine the vendor or product identity of the refractory concrete used in this flame deflector.

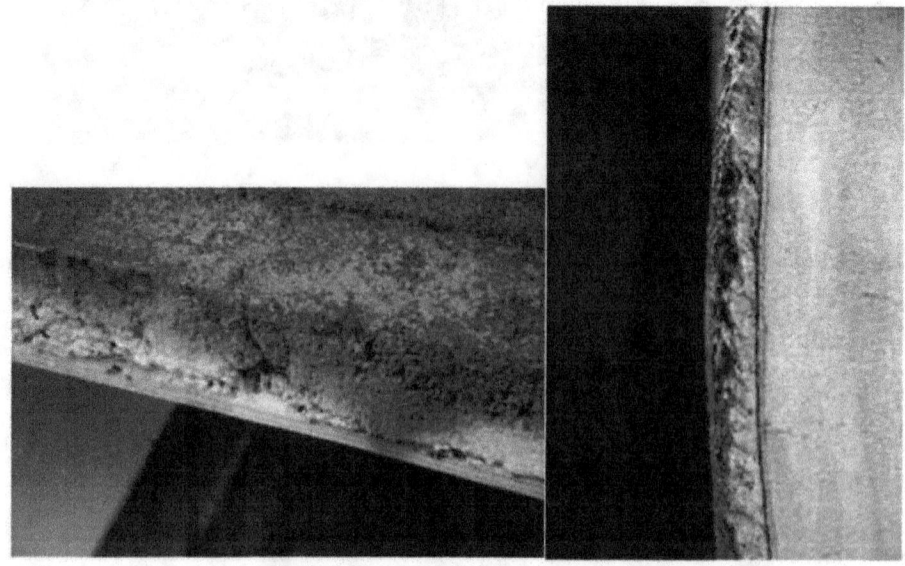

**Figure 20.   Refractory Concrete on Walls and Floor of
Stennis E-2 Cell 2 Flame Deflector**

5.1.4 E-3 Test Facility

The E-3 horizontal test cell can test motors up to 60 thousand foot pounds of thrust, and has support capabilities that include LOX, gaseous oxygen (GO_2), and GH_2. Cell 2 is a vertical test cell that is capable of testing engines that use LOX, hydrogen peroxide and HC propellants. Cell 2 can accommodate engines with thrust loads up to 25,000 foot pounds of thrust.[38]

Figure 21 shows the flame duct for the vertical E-3 Cell 2 test fixture. Numerous tests have been conducted at the vertical E-3 test stand. Tests have included small-scale combustion devices such as catalyst beds, to larger devices such as ablative thrust chambers and a flight-type engine.[39]

Testing, development and use of refractory material was evident during the site visit at the E-3 test cell. The original E-3 Cell 2 deflector was fabricated from a metal frame containing precast Fondu Fyre refractory concrete blocks. The precast blocks were secured to the flame deflector using steel angles bolted to the 1-1/2 inch thick steel side plates (Figure 21).[40]

Figure 21.   E-3 Vertical Test Cell Flame Duct at Stennis Space Center

Tests in this configuration show extensive damage to the steel angles, leading edges and fasteners. In the hot zone, the one-inch bolts securing the steel wall plates were either melted away or pulled out due to the thermal expansion (Figure 22). There is evidence that the refractory material melted from the impingement area and redeposited downstream (Figure 23).

---

[38] E-3 Test Facility. http://sscfreedom.ssc.nasa.gov/esd/ESDTestFacilitiesE3.asp. Last accessed on November 7, 2008.
[39] Jacks, T.E., and Beisler, M., "Expanding Hydrogen Peroxide Propulsion Test Capability at NASA's Stennis Space Center E-Complex," 39th AIAA/ASME/SAE/ASEE Joint Propulsion Conference and Exhibit, Huntsville, Alabama, July 20–23, 2003.
[40] E-mail correspondence from Nickey Raines, Stennis Space Center, October 7, 2008.

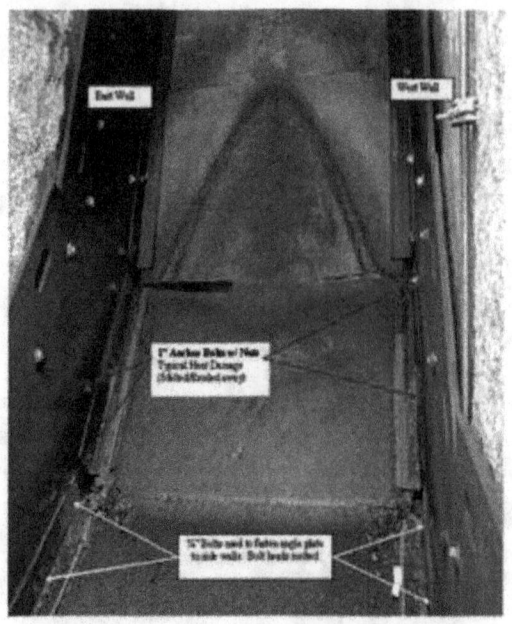

Figure 22.   E-3 Vertical Test Cell Flame Duct at Stennis Space Center – After Firing

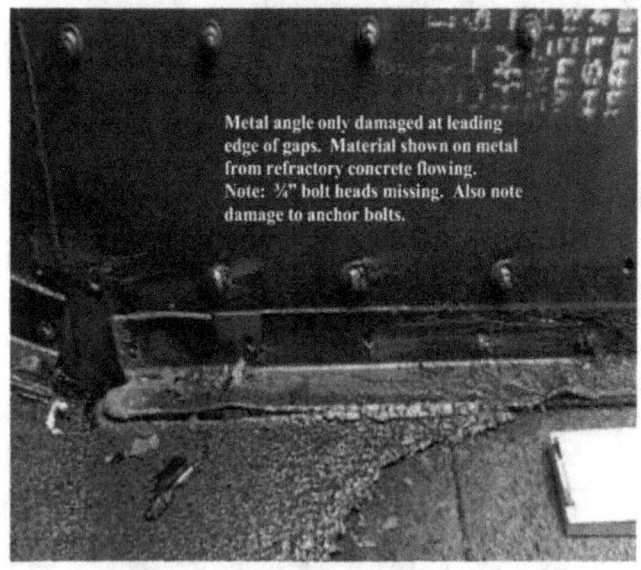

Figure 23.   Close-up view of Damage to E-3 Cell 2 Test Stand

Figure 24 shows a section of early Fondu Fyre that was used at the E-3 Cell 2 Test Stand. Historical test sections of Fondu Fyre were found throughout the E-3 facilities.

Figure 24.   **Fondu Fyre Blocks at the Stennis Space Center E-3 Location**

The blocks currently used in the E-3 Cell 2 Test Stand is a Fondag DG product. The blocks were cast in panels at the factory, and were built in a trapezoidal arrangement to overlap with each successive block. The blocks were cast at 300 °F, which had the result of both lowering the porosity and increasing the strength of the product.[41]

5.1.5   Bldg. 3300 Stennis Space Center

Building 3300 at Stennis Space Center contained remnants of components used for scale model testing of candidate refractory materials. Examples of these components are shown in Figure 25 and Figure 26.

Figure 25.   **Test Duct at Stennis Space Center**

---

[41] E-mail correspondence from Nickey Raines, Stennis Space Center, October 7, 2008.

Figure 26.   Refractory Concrete Apron at Stennis Space Center

These components were used as a part of a program designed to support the acquisition of data for baseline deflector design and refractory economical requirements. The program had four objectives:

- Establish ROM bounds on the extent of material loss and damage.
- Develop comparative data on the ability of various refractory materials to withstand the rocket plume environment.
- Develop engineering and scientific data characterizing surface and plume interaction phenomena.
- Evaluate, at model scale, the operational capability of the deflector.

Scaling of the test articles was driven by the availability of the Bates motor and propellant cartridges. The BATES motor (approximately 6000 lb thrust) was fired down the apron in a manner geometrically similar to that for a full-scale deflector. The plume deflector was designed in the configuration shown in Figure 27.[42]

Several literature sources discussed the design of the test assembly, as well testing of refractory concrete materials tested with the test fixture.[43,44,45,46]

---

[42] Douglas, F., Dawson, M.C., and Orlin, P.A., "ASRM Subscale Plume Deflector Testing," AIAA 17th Aerospace Ground Testing Conference, Nashville, Tennessee, July 6–8, 1992.

[43] Sauvel, J., "Static Test Defector for Ariane-6 SRM: Technical Design and Economic Choice," proceedings of the second European Conference on Progress in Space Transportation, May 22–24, 1989.

[44] Orlin, P.A., Dawson, D.M., and Bourgeois, S., "ASRM Plume Deflector Analysis," Sverdrup Technology, Inc., Report No. 3112-92-016, NASA/SSC, March 1992.

[45] Douglas, F.; "Subscale Test Measurement Data Accuracy," Stennis Space Center Report No. 3112-92-013, NASA/SSC, February 1992.

[46] "ASRM Subscale Deflector Test Report. Vol. 1: Rough Order of Magnitude (ROM) Testing," Sverdrup Technology, Inc., Report No. 311292-008 NASA/SSC, December 1991.

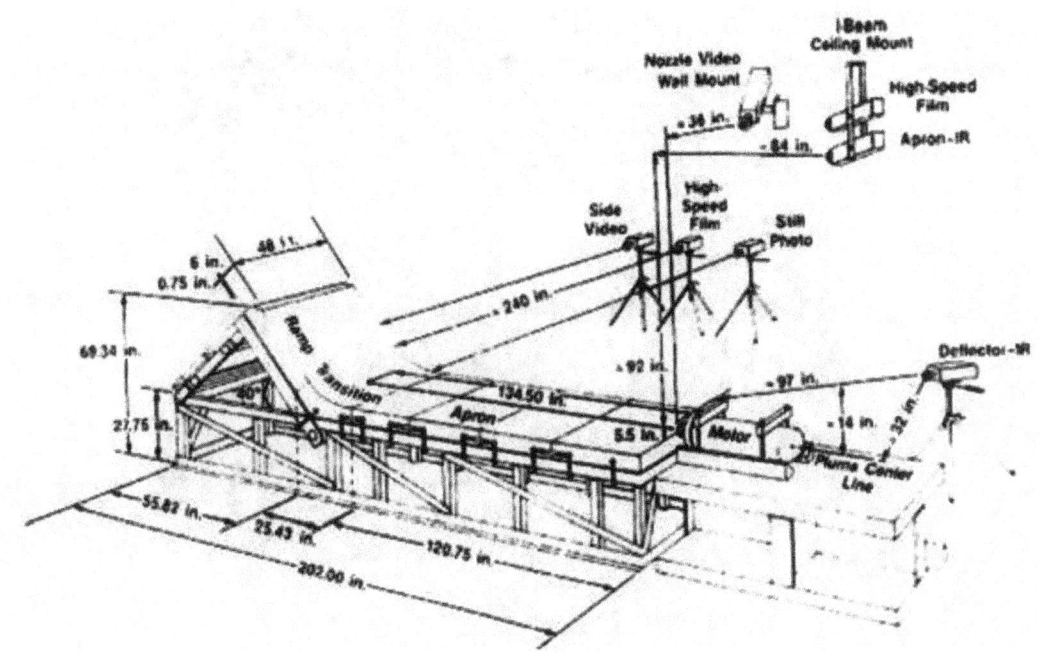

Figure 27. Stennis Plume Deflector Test Rig

5.2 Cape Canaveral Air Force Station Launch Facilities

The Cape Canaveral Air Force Station (CCAFS) is located on the east coast of Florida in Brevard County. The facility consists of over 15,000 acres of land, and is bounded by the Atlantic Ocean to the east and the Banana River to the west. CCAFS is part of the Air Force Eastern Range that includes administrative headquarters at Patrick Air Force Base, launch sites at Cape Canaveral, and downrange tracking facilities. The CCAFS area has been used to launch rockets by the United States government since 1949.[47]

5.2.1 Launch Complex 17

Launch Complex 17 (LC 17) was first built in 1956 for the THOR ballistic missile program, but later was used to launch probes to the Moon and planets, solar observatories, and weather satellites. LC 17 features two active Expendable Launch Vehicle (ELV) facilities. LC 17 began supporting launch operations in the late 1950s. The site was upgraded in the early 1960s to support modern ELVs. Delta II rockets have been launched from the site since the 1960s, though LC 17B was retrofitted in 1997 to support the Delta III program. The first launch of a Delta III system was on August 23, 2000.

An aerial view of the LC 17 is shown in its entirety in Figure 28.

---

[47] Historic American Engineering Record, Cape Canaveral Air Force Station, Launch Complex 17, U.S. Army Construction Engineering Research Laboratories, Report No. FL-8-5, December 1997.

Figure 28.   Aerial View of LC 17 [48]

LC 17A is shown to the right of the photograph. It utilizes an open flame trench (in front of the tower) to route the exhaust away from the Delta II rockets. LC 17B is shown to the left in the same photograph. In contrast to LC 17A, LC 17B was redesigned for the Delta III family of rockets. Changes included the addition of an enclosed flame trench, as well as flame ducts at each side for the SRBs.

LC 17 consists of two separate launch pads. These pads are designated LC 17A and LC 17B. Both launch pads include critical structures, such as the Fixed Umbilical Tower (FUT) and Mobile Support Tower (MST).[49]

Launch vehicles are initially assembled away from the launch complexes, and are only mated to the FUT after a significant portion of the rocket has been assembled. The FUT includes the launch pad and the Umbilical Tower (UT). After launch, the refurbishment team is allowed 3 to 5 days to prepare the complex for the next vehicle. Figure 29 shows the major components that make up LC 17 at CCAFS.

---

[48] Spaceflight Now | Delta Launch Report | Space Launch Complex 17.
http://www.spaceflightnow.com/delta/d327/070702slc17.html. Last accessed on November 11, 2008.
[49] Technology Evaluation for Environmental Risk Mitigation Principal Center.
http://www.teerm.nasa.gov/projects/DepaintingPollutionPreventionOpportunityAssessmentCCAFS.html. Last accessed on November 18, 2008.

Figure 29. Major Structures at LC 17[50]

Air Force Space Command (AFSPC) and NASA have launch facilities that reside in similar corrosive environments. Regardless of the corrosivity of the environment, all metals require periodic maintenance to guard against the insidious effects of corrosion, and thus ensure that structures meet or exceed design or performance life and ensure mission readiness requirements. As a consequence, LC 17 has been used as a platform for testing and evaluating various materials.[51]

5.2.1.1 LC 17A

A site visit was conducted at LC 17A on December 11, 2008.

The flame deflector at LC 17A uses a water deluge system to cool the rocket exhaust and dampen the acoustic levels. The flame deflector at LC 17A (viewed from the underside) is shown in Figure 30. The flame deflector was constructed using structural steel I-beams and reinforced concrete. Seams from the forms that were used for the concrete are visible in Figure 30. In contrast to other flame ducts, plate steel was not used on the underside of the flame duct refractory concrete.

---

[50] Depainting and Surface Preparation Pollution Prevention Opportunity Assessment for Cape Canaveral Air Force Station Space Launch Complex 17, Pads A & B, Final Report.
http://www.teerm.nasa.gov/reports/CCAFS_Dep_P2OA_Final_Report.pdf. Last accessed on November 19, 2008.
[51] Depainting and Surface Preparation Pollution Prevention Opportunity Assessment for Cape Canaveral Air Force Station Space Launch Complex 17, Pads A & B, Final Report.
http://www.teerm.nasa.gov/reports/CCAFS_Dep_P2OA_Final_Report.pdf. Last accessed on November 19, 2008.

Figure 30. Underside of LC 17A Flame Deflector

To protect the supporting concrete from the intense heat of the rocket exhaust, a layer of refractory concrete (Fondu Fyre) was originally applied for refractory protection. Throughout the launch program at LC 17, maintenance and repair has been required. To counteract the increased maintenance cycles, Martyte refractory has been integrated into the launch program.

Martyte was originally developed by Martin Marietta. It is a ceramic filled amine cured epoxy, and is delivered to the launch site as a three-part system.[52] It is a light colored material with an off-white appearance. Currently, all Fondu Fyre repairs at LC 17A are made with Martyte. Martyte is currently produced by Nitto Denko Automotive in Novi Michigan. Figure 31 shows the extent to which Martyte is applied to direct (and indirect) blast surfaces at LC 17A and the proximity to which the rocket nozzles are located in relation to the material.

---

[52] Sprayable Silicone Ablative Coating, GE Product Code CPC-1050, Post Launch Evaluation on Launch Complex.
[40] Cape Canaveral Air Force Station, Report No. 92-2150, http://corrosion.ksc.nasa.gov/92-2150.htm. Last accessed on November 19, 2008.

Figure 31. Flame Deflector at LC 17A

After launch, the flame deflector surface is "dust blasted" to remove soot and SRB residue. Once the postlaunch cleanup has been performed, the surface is inspected for spalling and ablation of the refractory materials. These repairs are made with Martyte, and are typically required after each launch. Figure 32 is an example of this requirement, and shows an ablated area of the flame duct that will use Martyte to cover the exposed Fondu Fyre substructure.

Evidence of prior Martyte patches is clearly visible in Figure 32. Individual Martyte patches, (such as those shown in Figure 32) are typically required after each launch.

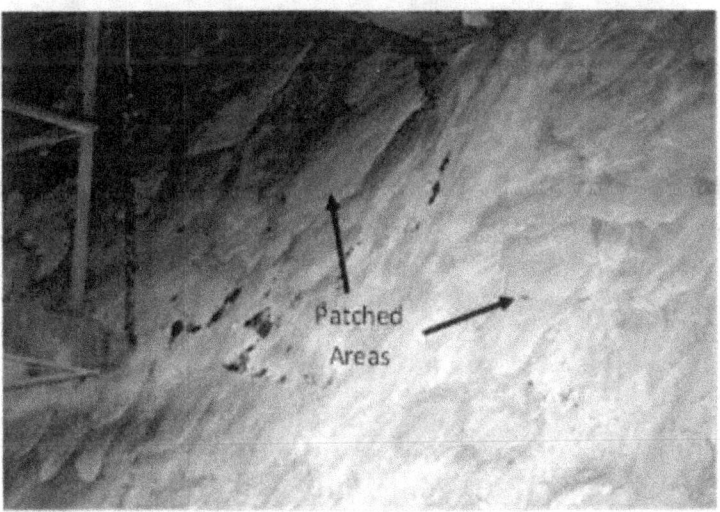

Figure 32. Fondu Fyre Patched Areas Using Martyte

5.2.1.2 LC 17B

LC 17B was refurbished for the Delta III rocket system, but only two Delta III rockets were ever launched from the facilities. Part of the refurbishment necessitated the construction of a flame duct by building two sidewalls to the J-turn deflector. This system uses a water deluge system for acoustic and thermal protection.

As opposed to LC 17A, LC 17B utilizes a covered flame duct tunnel to convey exhaust 150 feet downrange and away from the vehicle. Covered ducts were also added to each side of the launch pad, to safely route the exhaust from the Delta III graphite-epoxy SRB segments. [53]

The flame ducts at the side of LC 17B utilize Fondu Fyre for thermal protection (Figure 33). Fondu Fyre is applied at a depth of 3 to 4 inches. Little evidence of wear and abrasion was seen on the refractory material in this location. This is understandable since the side flame deflectors are not subjected to the direct impingement of rocket exhaust from the SRBs.

Figure 33.   Side Flame Duct at LC 17B

The side flame duct is positioned between the rocket, and an enclosed concrete tunnel that is used to further route the exhaust from the Delta rocket system (Figure 34). During launch, the side flame deflector is positioned immediately to the left of the concrete structure that is shown in Figure 34.

---

[53] Engblom, W.A., Weaver, M.A., and Nefo, E.D., "Numerical Study of Vehicle/Pad Configuration Effects on Launch Ignition Transients," 39th AIAA Aerospace Sciences Meeting and Exhibit, January 8–11, 2001.

Figure 34. Concrete Tunnel Adjacent to LC 17B

An inspection of the side tunnel supporting columns showed visible signs of erosion. Damage to the concrete supporting columns is shown in Figure 35. The supporting columns were built with construction grade (Portland cement) concrete. The damage was most likely caused by a combination of airborne particulates and thermal effects produced by the SRBs during launch.

Figure 35. Damage to Side Flame Tunnel Support Columns

In general, surfaces that are exposed to temperatures below 1500 F can use Portland cement mixes. Between 1000 and 1500 °F, aggregates in which quartz is the primary phase should be avoided. Above 1500 °F, ordinary Portland cement can be compromised by the heat.[54]

---

[54] Design Handbook for Protection of Launch Complexes from Solid Rocket Propellant, Report No. Martin-CR-66-11, Martin Marietta, March 1966.

In contrast to the open flame trench at LC 17A, the main flame trench at LC 17B is an enclosed duct that was designed to route the exhaust plume away from the rocket. Figure 36 shows the flame duct construction from the exhaust side of the tunnel. As indicated by the concrete apron in the center of the duct, LC 17B utilizes a water suppression system.

Figure 36.   Enclosed Flame Trench at LC 17B

The interior surface of the main flame trench largely consists of unprotected concrete. Exposed aggregate resulting from erosion to the concrete is shown in Figure 37.

Figure 37.   Abrasion to the Interior Surface of Main Flame Trench Walls

The flame deflector at LC 17B is constructed from a Fondu Fyre refractory material over a steel plate substructure. As shown in Figure 38, the main rocket impingement area in the center of the flame duct is protected by Martyte over Fondu Fyre refractory concrete.

The sidewalls of the flame duct (Figure 38) show large regions where Fond Fyre is still used for thermal protection. Once these regions spall and degrade, repairs are made with the alternate Martyte product Figure 39.

Figure 38.   Flame Duct at LC 17B

Figure 39.   Martyte Patch on Flame Deflector Side Wall

Repairs to the Fondu Fyre or Martyte are required after virtually every launch. Numerous patches from these operations are visible in the center of the flame duct in Figure 38.

The use of Martyte for thermal and ablative protection is not limited to the surface of the flame deflector. Critical launch components such as the nozzles to the water deluge system are also protected with the material (Figure 40), as well as bolt heads that are used to secure conduit enclosures (Figure 41).

Figure 40.   Martyte on Water Deluge System

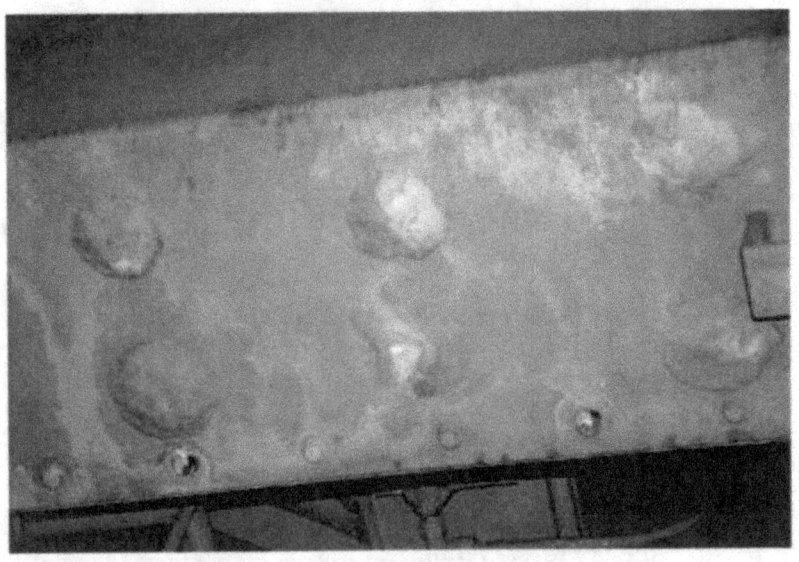

Figure 41.   Martyte-Protected Bolt Heads

The launch deck at LC 17B is positioned directly above the flame deflector. Figure 42 shows the proximity and distance (approximately 2 ft) of rocket nozzles in relation to the launch deck. As shown in the photograph, Martyte is used extensively on the launch deck of LC 17B. During launch, the majority of rocket exhaust is routed through ports. These ports are covered by metal doors as shown in the photograph in Figure 42. Understandably, direct rocket impingement occurs in areas in close proximity to these ports.

The use of Martyte for thermal protection on the launch deck is extensive. Interviews with LC 17 personnel, and a physical tour of the facility showed that the substructure in front of the line (in Figure 42) is typical construction grade concrete covered with Martyte refractory protection. The launch pad substructure in front of the demarcation line is typical structural steel that is protected by Martyte ranging from 1-1/2 inches thick near the base of the rocket down to 3/4 inch thick on the outer edges.

Figure 42.   LC 17B Launch Deck

Figure 43 shows a picture of one of the exhaust ports that had the protective metal cover removed for access. As shown in the figure, all direct impingement areas (perpendicular to the rocket blast) are covered with the thermal ablative. Surface areas that run parallel to the exhaust plume largely consist of structural steel that is coated for corrosion protection.

Figure 43.   LC 17B Exhaust Port on Launch Deck

Additional launch deck components are protected with a red thermal ablative coating (Figure 44). The pliability of the material would indicate that ablative is silicone-based. However, this supposition cannot be confirmed since the product is a Boeing proprietary formulation.

Figure 44.   **Ablative Coating on Tubing in Center of LC 17B Launch Deck**

5.2.2 Launch Complex 34

Launch Complex 34 (LC 34) was built in the late 1950s, and was constructed to launch the Saturn 1 and Saturn 1B family of rockets. Saturn 1B which is the larger of the two rockets, was 223 ft tall (with the Apollo capsule atop), and developed 1.6 million pounds of thrust at liftoff.[55]

From 1961 to 1968, four Saturn rockets were launched from LC 34. Unfortunately, the site is best known for the fire that killed astronauts Gus Grissom, Ed White, and Roger Chafee in a test for the Apollo 1 mission.[56]

An archived photograph shows the launch position of the flame deflector beneath the LC 34 pad during construction (Figure 45).

---

[55] John F. Kennedy Space Center - KSC Fact Sheets and Information Summaries. http://www-pao.ksc.nasa.gov/nasafact/count1.htm. Last accessed on December 17, 2008.

[56] Pad 34. http://www.floridatoday.com/apps/pbcs.dll/article?AID=/99999999/NEWS09/60312017/1007. Last accessed on January 5, 2009.

Figure 45. LC 34 Flame Deflector and Launch Pad [57]

The original flame deflectors remain at the historical LC 34 launch site. These items were inspected on December 11, 2008.

Prior to the inspection, several NASA personnel had indicated that the flame deflectors were all metal (water-deluge-cooled) units. An inspection of the facilities and flame ducts (Figure 46) showed that a thermal ablative was indeed integrated into the design as a key component. A visual inspection of the flame deflectors indicated that the depth of the refractory material varied quite extensively, though it was thickest in the areas where the floor and walls of the flame deflector meet.

Figure 46. Flame Deflector at LC 34 on December 11, 2008

---

[57] Moonport, CH2-4; http://www.hq.nasa.gov/office/pao/History/SP-4204/ch2-4.html. Last accessed on December 17, 2008.

Figure 47 shows a magnified section of the refractory material in the center of the flame deflector floor.

Figure 47.   Refractory Material at LC 34

As of the writing of this report, the authors were unable to determine the product identity, nor the vendor for the refractory material. This is understandable considering the significant age of the structure and the number of years that the item has remained as a historical remnant of the Apollo era.

5.2.3   Launch Complex 36

The military constructed Launch Complex 36 (LC 36) for the Atlas rocket program. A second pad was added in the mid 1960s.

NASA operated the launch complex for most of its service life and launched scientific missions that included:

- Surveyor spacecraft that landed on the moon in advance of the manned Apollo missions.

- The Mariner robotic probes for Mars and Venus.

- Several Pioneer long-distance probes.

- Numerous commercial and military satellites.

LC 36 is no longer used for launch and was decommissioned from use in 2005.[58] Structural steel from the launch pad was removed from the site and used as an artificial reef in the Atlantic Ocean. Based upon a Lockheed-Martin article pertaining to this operation, it was determined that

---

[58] Pads 36A & 36B. http://www.floridatoday.com/apps/pbcs.dll/article?AID=/99999999/NEWS09/60312018/1007. Last accessed on January 6, 2009.

Martyte was a key refractory material used on the flame deflector along with Fondu Fyre refractory concrete.

Information regarding the refractory concrete used at the site was unavailable but a search of the literature revealed that Martyte was used as a component of the flame deflectors.[59]

5.2.4 Launch Complex 37

The construction of Launch Complex 37 (LC 37) began in 1959 to support the Saturn IB rocket program. LC 37B was used for unmanned Saturn I and IB flights through the mid 1960s.

LC 37 was completely refurbished in the late 1990s. Completion of the project occurred in 2000 with funding financing from the Florida Space Authority. Currently, the facility is used by Boeing to launch the Delta 4 series of rockets.

The first heavy version of the Delta 4 flew from Pad 37 in December 2004.[60] A prelaunch photograph of the vehicle is shown in Figure 48.

Figure 48.   Delta IV Heavy at Launch Complex 37 [61]

---

[59] Lockheed Martin New Horizons. http://www.lockheedmartin.com/data/assets/10006.pdf. Last accessed on January 5, 2008.
[60] Pad 37. http://www.floridatoday.com/apps/pbcs.dll/article?AID=/99999999/NEWS09/60312023/1007. Last accessed on January 5, 2008.
[61] Delta 4 Heavy. http://www.astronautix.com/graphics/d/delta4h.jpg. Last accessed on January 1, 2009.

A search of the literature indicated that Fondag DG is the principal refractory concrete used at LC 37.[62] Based upon the referenced source, Atlantic Firebrick of Jacksonville, Florida, was contacted for further information.

Atlantic Firebrick confirmed that Fondag DG was the principal refractory material in use on the flame deflector at LC 37. Kruzite GR Plus, which is a gunnable refractory concrete, was also used at LC 37. A desirable quality of Kruzite GR Plus centers upon its ability to adhere to overhead structures better than Fondag DG. Therefore, less rebound is experienced with this product.[63]

Further information regarding refractory materials in the flame trench at LC 37 was unavailable at the time this report was finished. Because of the proprietary and secretive nature of the Delta IV launches, flame trench materials and designs were closely guarded by LC 37 launch personnel.

5.2.5 Launch Complex 40

Launch Complex 40 (LC 40) was used by the United States Air Force to launch Titan III and Titan IV rocket systems until April 2005. The Titan III system was principally used to launch commercial satellites.[64]

LC 40 utilizes a water deluge system for thermal cooling and sound suppression. Discussion with Kerneos Aluminate Technologies personnel indicated that the principal refractory concrete at LC 40 is the Fondag DG product.[65] Additional information on refractory materials at the launch site was obtained from NASA Test Report No. 92-2150.[66] This test report illustrated the use and performance of the General Electric (CPC-1050) Silicone ablative coating on structural steel. The performance this product was rated as excellent after it was exposed to Titan 34D launch conditions.

The same report noted that Martyte was a product in use at the facility, and made specific mention of the standard practice of protecting water deluge nozzles with the material. The same report commented on possible incompatibility issues between the silicone ablative and epoxy Martyte refractory. After the launch of a Titan 34D rocket, Martyte refractory sections (that were applied over the silicone ablative) delaminated from the underlying surface.

On April 25, 2007, LC 40 was leased to SpaceX to launch their Falcon 9 rocket system. NASA selected the SpaceX Falcon 9 launch vehicle and Dragon spacecraft for the International Space

---

[62] Fitzgerald, M.W., Talley, J., and Alt, C.W., "Calcium Aluminate Technology and Its Application in Refractory Concrete," Shotcrete, Summer 2002.
[63] Telephone interview with Doug Goddard, Atlantic Firebrick, January 6, 2008.
[64] Pads 40 & 41. http://www.floridatoday.com/apps/pbcs.dll/article?AID=/99999999/NEWS09/60312021/1007. Last accessed on January 9, 2009.
[65] Telephone interview with Greg Wallace, Kerneos Aluminates, January 6, 2009.
[66] "Sprayable Silicone Ablative Coating, GE Product Code CPC-1050, Post Launch Evaluation on Launch Complex," NASA Report No. 92-2150, November 23, 1992.

Station Cargo Resupply Services (CRS) contract award. Under the contract, SpaceX will be responsible for 12 flights between 2010 and 2015.[67]

According to the Falcon 9 data sheet, the total thrust from the SpaceX Merlin engines is estimated at 1.125 million foot pounds.[68] Estimated thrust for the Falcon 9 Heavy is estimated to be 3.375 million foot pounds.[69]

According to Kerneos Aluminate Technologies personnel, SpaceX continues to use Fondag DG as the principal refractory material in the flame trench at LC 40.[70]

5.2.6 Launch Complex 41

Launch Complex 41 (LC 41) was initially built to support the Titan launch program. The first Titan 3C was launched from this facility in December 1965. The last Titan flight to occur was in April 1999.

To accommodate the new Evolved Expendable Launch Vehicle (EELV) program, LC 41 was redesigned and rebuilt for the new Atlas V series of rockets. Atlas V is an expendable launch vehicle that is built through a joint alliance between Lockheed–Martin and Boeing under the name United Launch Alliance.[71]

Preparation for the new launch system necessitated the removal of the Mobile Service Structure and construction of a Vehicle Integration Facility. The Vehicle Integration Facility, was finished in 2000, and a new Control Center was completed in 2001.

The Atlas system has launched numerous commercial satellites from LC 41. Other payloads have included NASA's Mars Reconnaissance Orbiter and the Pluto New Horizons probe.[72]

LC 41 Refractory materials were investigated for this report. The refractory materials at LC 41 must be able to withstand the deleterious effects produced by the kerosene\LOX fueled rocket motor and up to five SRBs.[73] To counteract the heat and acoustics of the rocket system, a water deluge is activated three seconds into the launch.

---

[67] Space Exploration Technologies Corporation – SpaceX. http://www.spacex.com/. Last accessed on January 6, 2009.
[68] Falcon 9 Launch Vehicle Data Sheet. http://www.spacex.com/Falcon9DataSheet.pdf. Last accessed on January 12, 2009.
[69] Space Exploration Technologies Corporation - Falcon 9 Heavy. http://www.spacex.com/falcon9_heavy.php. Last accessed on January 12, 2009.
[70] Telephone interview with Greg Wallace; Kerneos Aluminates; January 6, 2009.
[71] Pads 40 & 41. http://www.floridatoday.com/apps/pbcs.dll/article?AID=/99999999/NEWS09/60312021/1007. Last accessed on January 9, 2009.
[72] Pads 40 & 41. http://www.floridatoday.com/apps/pbcs.dll/article?AID=/99999999/NEWS09/60312021/1007. Last accessed on January 9, 2009.
[73] Atlas V data sheet. http://www.ulalaunch.com/docs/product_sheet/AtlasProductSheetFINAL.pdf. Last accessed on January 9, 2009.

The Atlas V flame bucket was rebuilt specifically for the Atlas V system. A review of the facilities indicated that the flame deflector utilizes Fondu Fyre refractory concrete, over a reinforced concrete substructure. To further protect the flame deflector, a Havaflex phenolic ablative coating is applied to a depth of approximately 1 inch.

Havaflex is a trowelable and sprayable phenolic ablative coating that was originally developed for the U.S. Navy to protect decks, bulkheads, and shipboard launchings systems. The product was designed for high temperature (up to 5000 °F.) and high gas velocity (up to MACH 3.0) missile launch operations.[74]

Additional refractory protection is afforded with the use of the Martyte epoxy ablative on instrumentation in the flame trench at LC 41. In contrast, a silicone ablative coating is used to protect connections from excessive heat, erosion and ablation..

Because of the proprietary and secretive nature of the launch activities at LC 41, photographs of the launch pad and flame duct system were not allowed, and therefore could not be included in this report.

5.3 Vandenberg Air Force Base: Launch Complex 6

With the advent of the missile age in the 1950s, Vandenberg AFB (Camp Cooke at the time) was chosen as America's first combat ready missile base. This decision was based upon size, remoteness from heavily populated areas, desirable climate features and coastal location. The proximity to the ocean is also beneficial, since it provides a margin of safety by avoiding flights over populated areas.

On December 16, 1958, Vandenberg successfully launched its first missile, a Thor intermediate range ballistic missile (IRBM). Over the years, Vandenberg has been used as a location to launch unmanned satellites. As of November 2005, a total of 1858 orbital and ballistic missiles have been launched from Vandenberg AFB facilities.[75]

Launch Complex 6 (LC 6) was originally designed for the Titan III program, and eventually was transformed through a 4-billion-dollar modification for the Space Shuttle. For reasons beyond the scope of this report, Space Shuttle operations were centralized to the Kennedy Space Center.[76]

In the late 1990s, Boeing modified LC 6 for the Delta IV series of rockets. To date, the modified facility has been used to launch the Delta IV Medium, and Delta IV Medium + series of rockets.[77]

---

[74] Havaflex T.A.-117 A Trowelable Ablative Material. http://new.ametek.com/content-manager/files/HAV//Havaflex1.pdf. Last accessed on January 9, 2008.
[75] Vandenberg Air Force Base - Fact Sheet; Vandenberg Air Force Base - Fact Sheet. http://www.vandenberg.af.mil/library/factsheets/factsheet.asp?id=4606. Last accessed on January 9, 2009.
[76] Space Vehicles: History Office. http://www.vandenberg.af.mil/library/factsheets/factsheet.asp?id=4606. Last accessed on January 9, 2008.
[77] Boeing: Boeing Completes First Delta IV West Coast Launch. http://www.boeing.com/ids/network_space/news/2006/q2/060627b_nr.html. Last accessed on January 9, 2009.

Figure 49 shows the assembly building (in the foreground) for the Delta IV rockets at LC 6. The water deluge system that is used for launch operations is shown in the background of the same figure. The deluge water is recovered after each launch, recycled, and discharged in accordance with applicable industrial wastewater permits and regulations.[78]

Figure 49.   LC 6 at Vandenberg Air Force Base[79]

Kerneos Aluminate Technologies was contacted to ascertain the extent to which their product was used at Vandenberg Air Force Base. Kerneos personnel indicated that Fondag DG was used as the primary refractory concrete at LC 6.[80] Further discussion indicated that other launch facilities at Vandenberg only utilized construction grade (Portland cement) concrete. Because of the proprietary and secretive nature of launches at Vandenberg, the authors of this report were unable to verify this statement.

5.4   Palm Beach County Fire Training Facility

The Palm Beach County Fire Rescue Administration and Training complex was designed to provide training simulations designed to meet the needs of the fire service.

The 40-acre complex is located on Pike Road, near the entrance to the Florida Turnpike, and houses administrative offices, training classrooms, practical training areas and an apparatus and support building. The concrete burn building (under construction) is designed to simulate a two-story 4200 ft^2 residential structure.[81] An exterior view of the facility is shown in Figure 50.

---

[78] Final Supplemental Environmental Impact Statement (FSEIS) for the Evolved Expendable launch Vehicle (EELV). http://www.faa.gov/about/office_org/headquarters_offices/ast/licenses_permits/media/SEISROD2000-2000.pdf. Last accessed on January 9, 2009.

[79] Boeing: Multimedia - Image Gallery - Delta IV Launch Site - Space Launch Complex 6. http://www.boeing.com/companyoffices/gallery/images/space/delta_iv/d4_slc6_08.html. Last accessed on January 9, 2009.

[80] Telephone interview with Greg Wallace, Kerneos Aluminates, January 6, 2009.

[81] Palm Beach County Fire Rescue - Herman W. Brice Training Complex Construction Information. http://www.pbcfr.org/training_construction.asp. Last accessed on November 4, 2008.

The late construction timeframe provided a unique opportunity to inspect the refractory system and its installation.

Figure 50.   Palm Beach County Fire Rescue Training Facility

The refractory system used in the Palm Beach County Fire Rescue Training Facility was provided and installed by High Temperature Linings in Fairfax, Virginia.[82] The trade name of the company's refractory product is System 203. Per the manufacturer, System 203 is able to withstand repeated high temperatures and extreme thermal shock created during fire training exercises.

Each System 203 Firetile is individually anchored to the structure with a stainless steel anchoring system (Figure 51). Coupled with the interlocking feature of the tiles (Figure 52), the lining is allowed to "float" with temperature fluctuations.[83]

---

[82] High Temperature Linings, 606 Chesapeake Drive, Suite C, PO Box 1240, Whitestone, Virginia 22578, (800) 411-6313.
[83] High Temperature Linings. http://www.firetrain.com/system203/system203.html. Last accessed on January 9, 2009.

Figure 51. System 203 Tiles at the Palm Beach County Fire Training Center

Per the manufacturer, the system is designed to last for 10 to 20 years (for Fire Training Buildings) with only minimal maintenance. The system is:[84]

- **Designed for maximum temperatures between 1600 and 2200 °F.**

- **Resistant to thermal shock.** The lining is designed to "float" on top of a calcium silicate insulation board.

- **Suitable for all types of buildings.** The lining anchors to concrete masonry units. Alternatively, it may be installed to a framework of channels anchored to structural steel if proper ventilation and waterproofing measures are taken.

- **Designed to keep water and steam out (for interior fire training buildings).** The ship-lapped design protects the insulation and structure from water, steam, and heat damage.

The interlocking design of the System 203 tiles is shown in Figure 52.

---

[84] High Temperature Linings. http://www.firetrain.com/system203/system203.html. Last accessed on January 9, 2009.

Figure 52.   Interlocking System 203 Tiles

The system 203 tiles are produced from a proprietary formulation that contains stainless steel fibers for strength. The insulating (and structural concrete protecting) properties of the system result from the use of a calcium silicate backboard underneath the temperature resistant tiles.[85]

The System 203 product is marketed as a rugged product that is capable of withstanding the abuse of high-pressure streams of water.[86] Unfortunately, the use of System 203 in this manner is not comparable to the launch pad environments at KSC. These environmental conditions include extreme erosion, ablation, and greatly fluctuating temperatures. Furthermore, the complexity of the system, while useful for the interior of a building, may present problems for postlaunch repairs and maintenance.

6   CONCLUSIONS

Refractory concrete is used to protect National Aeronautics and Space Administration (NASA) launch structures from elevated temperatures, ablation, and erosion. The only refractory material qualified for use at Launch Complex 39A (LC 39A) and Launch Complex 39B (LC 39B) is Fondu Fyre WA-1G which is supplied by the Pryor Giggey Co. The material was developed solely for NASA in the 1960s.

Refractory concrete at LC 39A and LC 39B have become susceptible to failure, resulting in large sections of the materials breaking away from the base structure. During launch, these sections become high speed projectiles that jeopardize the safety of KSC personnel, and have the potential to damage ground support equipment and the Space Shuttles.

---

[85] System 203. http://www.firetrain.com/PDFs/system%20203.pdf. Last accessed on January 9, 2009.
[86] High Temperature Linings. http://www.firetrain.com/system203/system203.html. Last accessed on January 9, 2009.

A review of the current specification and requirements for refractory materials indicates that the test methods and qualification criteria are not well defined. Consequently, the only refractory product qualified for use at KSC may not have the material properties necessary to survive extended exposure to Florida coastal environments and the severe launch conditions exhibited by the Space Shuttle. As a result, better performing refractory materials may be available for use at KSC.

A literature search was conducted to ascertain the different categories of refractory materials that are available for the protection at KSC's launch pads. The classes of materials were categorized as follows:

- **Firebrick**
- **Refractory Concrete**
- **Silicone and Epoxy Ablatives**

Based upon this information, a literature survey was conducted to locate industries that had refractory requirements that were similar to NASA's. Based upon this survey, site visits, and interviews with pertinent industry personnel and refractory vendors were conducted. The following list summarizes the site visits, rockets, test site load capacities, and the materials used to protect the flame deflectors.

- **Stennis Space Center**
    - A-2 Test Facility
        - Metal flame deflector
        - 1.1 million foot pound capacity
        - Liquid propellants
    - B1 and B2 Test Facilities
        - Metal flame deflector
        - 11 million foot pound capacity (each structure)
        - Liquid propellants
    - E-2 Test Facility
        - Unknown refractory concrete flame deflector
        - 100,000 foot pound capacity
        - Liquid and RP-1 propellants
    - E-3 Test Facility
        - Precast Fondu Fyre flame deflector (original)
        - Precast Fondag DG flame deflector (current)
        - 25,000 foot pound capacity
        - Liquid, hydrogen peroxide and hydrocarbon fuels
    - Bldg 3300
        - Remnants of multiple refractory tests (from Edwards AFB/JPL testing)
        - 6000 foot pound thrust for testing
        - Bates solid rocket motor

- **Cape Canaveral Air Force Station**
  - Launch Complex 17
    - Fondu Fyre WA-1G and Martyte refractory material along with a Boeing proprietary ablative
    - Delta II and Delta III
    - Liquid and solid fuels
  - Launch Complex 34
    - Unknown refractory concrete flame deflector
    - Saturn 1 and Saturn 1B Rocket Systems
    - Liquid fuels
  - Launch Complex 36
    - Fondu Fyre and Martyte refractory materials
    - Atlas I, Atlas II and Atlas III
    - Liquid and solid fuels
  - Launch Complex 37
    - Fondag refractory concrete
    - Delta IV
    - Liquid and solid fuels
  - Launch Complex 40
    - Fondag refractory concrete, Martyte and GE silicone ablative
    - Titan III, Titan IV, Falcon 1 and Falcon 9
    - Liquid and solid fuels
  - Launch Complex 41
    - Fondu Fyre WA-1G refractory concrete, Havaflex phenolic, and various silicone ablatives
    - Atlas V
    - Liquid and solid fuels
- **Vandenberg Air Force Base**
  - Launch Complex 6
    - Fondag refractory concrete
    - Delta IV
    - Liquid and solid fuels
- **Palm Beach County Fire Department**
  - Fire Training Facility
    - Proprietary precast refractory cement and calcium silicate insulating backboard system
    - Hydrocarbon fuels

As a result of the site visits and interviews, a series of products for launch applications were found.

Firebrick, while historically used near flame deflectors at NASA launch sites, was not found at any of the locales investigated in this report. Product and labor costs associated with the installation of the materials were cited as the driving factors for its lack of use.

Refractory concrete was used at numerous launch locations. Currently used products include Fondu Fyre WA-1G which is produced by the Prior Giggey Co., and Fondag DG, which is supplied by Kerneos Aluminate Technologies. Both versions are gunnable, and as a result, benefit from reduced labor costs associated with the application of the product. Kruzite GR Plus is another refractory concrete that is noteworthy. Kruzite GR Plus is currently being used and was recommended for use in overhead flame duct locations due to its better adhesion (less rebound) properties.

The direct impingement areas (of the flame deflectors) were often found to be protected by a ceramic filled epoxy called Martyte. Martyte was often used to replace and protect refractory concrete that had deteriorated. Furthermore, structural steel (in direct impingement areas) was often protected by the product.

Havaflex is a phenolic ablative that is produced by Ametek Chemical Products. It can be either troweled or sprayed as required, and is used in areas that are subject to direct rocket exhaust.

Various silicone ablative materials were used outside direct blast areas. These coatings included the General Electric GE 3404 ablative, as well other proprietary formulations from other manufacturers and aerospace companies. These products are often used to protect structural steel, launch pad tubing, and connectors for launch pad instrumentation.

None of the products in this trade study can be considered a panacea for LC 39A and LC 39B. Fondag DG, while inexpensive, was often top coated with Martyte for repair or additional thermal protection.

Martyte is costly and difficult to apply. Furthermore, incompatibilities between Martyte and the silicone ablatives may be of concern.

Havaflex is a phenolic ablative material that is easy to apply; unfortunately, it is costly and requires frequent replacement.

The silicone ablatives are inexpensive, easy to apply, and perform well outside of direct rocket impingement areas. When used in locations subject to direct rocket exhaust, the performance of the coating is exceeded by refractory concrete and the epoxy alternatives.

As discussed in this report, the site inspections and interviews with launch complex personnel revealed that the number of refractory products routinely used for launch applications is extremely limited. In a separate report (WBS 5.2.2.1.6 COTS vs. Refractory Materials Requirements for Flame Deflector Trade Study) that is in preparation for NASA's Technology Development Program, over 800 products were identified and their material properties were documented. The cold modulus of rupture, hot modulus of rupture, thermal conductivity, cold crushing strength, application method and (manufacturers reported) recommended service temperature were used to assess the product for flame deflector use. Once ranked by these reported specifications, over 114 products were determined to be viable candidates to be considered for physical testing and evaluation under NASA's stringent requirements.[87]

This report summarizes the ablative materials that were found at industries with refractory requirements that are similar to NASA's. The refractory products, individually or in combination, may be considered for use at LC 39A and LC 39B provided the appropriate testing requirements and specifications are met. Unfortunately, due to the extreme conditions experienced at LC 39 and unique nature of the launching rockets, none of the products promise to be the drop in replacement that will solve the delamination and spalling of refractory materials that occurs at NASA's launch sites. Further testing of promising candidates needs to be addressed.

---

[87] Calle, L.M., Hintze, P.E., Parlier, C., Curran, J.P., Kolody, M.R., Perusich, S.A., Whitten, M.C., Trejo, D., Zidek, J., and Coffman, B., "WBS 5.2.2.1.6 COTS vs. Refractory Materials Requirements for Flame Deflector Trade Study," National Aeronautics and Space Administration, in preparation.

APPENDIX A.   PRODUCT DATA SHEETS

TECHNICAL DATA

FONDU FYRE® WA-1G is an erosion resistant gunning concrete designed for exposure to rocket engine exhaust for launch or test stand applications.

Maximum Recommended Use Temperature:	Sustained temp. 2200°F
Material Required:	120 - 125 pcf
Thermal Conductivity (Btu·in/h·ft²·°F):	5.0 @ 1000°F, 5.2 @ 1500°F
Cold Crushing Strength (psi @ 7 days)	4,000 - 6,000
Porosity (%)	16
Coefficient Thermal Expansion (x10⁻⁶/In/In/°F)	4.5

Chemical Analysis

Al_2O_3	26.5%
SiO_2	34.0%
Fe_2O_3	11.5%
TiO_2	0.5%

Packaging:	55# and bulk bags
Date Last Revised:	9/89

Chehalis, WA • Anniston, AL
1-800-446-8769
All data subject to reasonable deviations and not to be used for specification purposes

Product data sheet

Reference PDS-US-FDG-8/06

Fondag® DG

Updated 8/23/2006

1 General Characteristics

Composed entirely of calcium aluminates, FONDAG® DG is a pre-packaged ready to use high strength dry gunite.

FONDAG® DG is a concrete that can be used in severe duty industrial concrete applications that may require refractory performance.

FONDAG® DG is recommended for concrete applications requiring rapid hardening properties, resistance to abrasion and mechanical shock, resistance to chemical attack and exposure to refractory temperatures < 2000° F.

FONDAG® DG does not release calcium hydroxide as a hydration product. This imparts good refractoriness, chemical resistance and eliminates the major cause of efflorescence.

FONDAG® DG is a very dark gray color. Colorimetry data is available on request.

FONDAG® DG does not contain any additives.

FONDAG® DG does not contain crystalline silica.

2 Specifications

FONDAG® DG produced and distributed in North America adheres to the following specifications:

Sieve Analysis

	Min (%)	Max (%)
# 4 (4.75mm)	0	0
# 8 (2.36 mm)	0	12
# 16 (1.18 mm)	23	46
# 30 (600 µm)	40	65
# 50 (300 µm)	50	72
# 100 (150 µm)	57	76
# 200 (75 µm)	61	79
Pan	20	40

Mortar Properties (using 14% water)

▫ Penetrometer Final Set: 2 – 5 hours

For detailed test procedures, please contact a Kerneos Technical or Quality Manager.

3 Additional Physical properties

Kerneos Inc.
1316 Priority Lane Chesapeake, VA 23324
Phone (757) 284-3200 - FAX (757) 284-3300

Page 1 (2)

ISO 9001

Product data sheet

Reference PDS-US-FDG-8/06

4. Packaging & Shelf Life

FONDAG® DG is available palletized in 65 lb bags.

FONDAG® DG packaging is designed to protect it from humidity. However, as with all prepackaged concretes, it is recommended that FONDAG® DG not be placed outdoors or in direct contact with the ground. When correctly stored in dry conditions, the properties of FONDAG® DG will remain within specification limit for at least 6 months. In most cases, its properties will be retained for over a year.

KERNEOS LIMITED WARRANTY
Kerneos warrants that this product, at the time of shipment, conforms to the Specifications set forth in section 2 of this Product Data Sheet. All other information provided in this Product Data Sheet is for guidance only. ALL OTHER WARRANTIES, INCLUDING WITHOUT LIMITATION THE WARRANTIES OF MERCHANTABILITY AND FITNESS FOR A PARTICULAR PURPOSE, ARE EXCLUDED. Kerneos' sole obligation and the sole and exclusive remedy under this limited warranty shall be the replacement of any nonconforming product, or, at Kerneos' option, the refund of the purchase price. No warranty is given for any technical advice or recommendations provided by Kerneos. Buyer waives all claims under this limited warranty unless it has given written notice of nonconformity within 30 days of delivery.

ISO 9001

Product Data

10/08: 5885

KRUZITE® GR PLUS

Physical Properties: (Typical)	English Units	SI Units
Description: 70% Alumina Gunning Castable		
Maximum Temperature	3200°F	1760°C
	lb/ft³	g/cm³
Material Required	145	2.32
Bulk Density		
After 220°F (105°C)	149	2.39
After 1500°F (815°C)	145	2.32
After 2500°F (1371°C)	142	2.27
After 2910°F (1600°C)	142	2.27
Water Required for Premixing	See Mixing & Using Instructions	
Permanent Linear Change, %		
After 1500°F (815°C)	−0.2	
After 2500°F (1371°C)	0.0	
After 2910°F (1600°C)	+0.9	
Modulus of Rupture	lb/in²	MPa
After 220°F (105°C)	1500	10.3
After 1500°F (815°C)	1000	6.9
After 2500°F (1371°C)	1700	11.7
After 2910°F (1600°C)	2000	13.8
Hot Modulus of Rupture		
At 2500°F (1371°C)	750	5.2
Cold Crushing Strength		
After 220°F (105°C)	4800	33.1
After 1500°F (815°C)	5800	40.0
After 2500°F (1371°C)	7100	49.0
After 2910°F (1600°C)	4600	31.7
Thermal Conductivity		
At a Mean Temperature of	Btu·in/hr·ft²·°F	W/m·°C
400°F (205°C)	9.8	1.41
800°F (425°C)	9.1	1.31
1200°F (650°C)	8.6	1.24
1600°F (870°C)	8.3	1.20
2000°F (1095°C)	8.2	1.18
2400°F (1315°C)	8.2	1.18

(Continued)

Product Data

KRUZITE® GR PLUS (Continued)

Chemical Analysis: (Approximate)
(Calcined Basis)

Silica	(SiO_2)	26.2%
Alumina	(Al_2O_3)	67.6
Titania	(TiO_2)	2.1
Iron Oxide	(Fe_2O_3)	1.1
Lime	(CaO)	2.8
Alkalies	(Na_2O+K_2O)	0.1

Description: KRUZITE GR PLUS is a 70% alumina, dense, 3200°F maximum service temperature gunning castable. It exhibits excellent strengths and high densities. Typical applications are: iron torpedo ladle and hot metal mixer gunned maintenance linings, general steel mill maintenance material, air heater linings, boiler repairs, soaking pit bottoms, high temperature ductwork, and high temperature stack linings.

The test data shown are based on average results on production samples and are subject to normal variation on individual tests. The test data cannot be taken as minimum or maximum values for specification purposes. ASTM test procedures used when applicable.

10/3/08 Dev.

HAVAFLEX T.A.-117
A TROWELABLE ABLATIVE MATERIAL

Product Description
Havaflex T.A.-117 is a two-component modified phenolic ablative coating and adhesive. This material was originally developed for use by the U.S. Navy to provide a flexible trowelable coating material to protect decks, bulkheads, and shipboard launchings systems from the extremely high temperatures (up to 5000°F.) and high gas velocities (up to MACH 3.0) present during missile launch operations.

Technical Data
Properties
Havaflex T.A.-117 is flexible, an excellent thermal insulator, will not support combustion, and is resistant to many chemical environments, including acid and salt solutions. The material has a pot life of 30 to 40 minutes at room temperature and cures within 24 to 36 hours. The shelf life is approximately one year when stored at 40°F or below. Havaflex TA-117 will adhere to almost all types of surfaces (free of moisture, dirt, or grease) without the need of a primer. Havaflex T.A.-117 can be machined, sawed, planed, drilled, ground or sanded, spun molded and can be applied vertically in ¼ inch layers to desired thickness without sagging.

Physical Properties

Tensile Strength (psi)............625	Tensile Lap Shear (psi).....625 (ASTM D 638) (ASTM D 1002)
Tensile Elongation, %...........12.8	Water Absorption, % (ASTM D 638) (96 hrs. at 65-75°F).........0.67
Density (lbs/ft3)..................81.2	Flame Spread Index.........143 (ASTM E 162)
Specific Gravity..................1.30	Coverage (1/4" thick).......6.2ft2 per kit
Thermal Conductivity...........3.56 (Btu/ft2/°F./hr/in of path)	Coefficient of Thermal Expansion (in/in/°Cx10-6) perpendicular 2.90
Shore Hardness..................90 Type A	(in/in/°Cx10-6) parallel 7.00
Specific Heat.....................0.206 (ASTM E 1269-94) (Btu/lb/°F)	

Packaging
Havaflex T.A.-117 is available in one-gallon kits, and is packaged in plastic containers with Part A and Part B in separate containers. The two components can be easily mixed together by using a ½ inch electric drill with a stainless steel Jiffy mixing blade.
Product Bulletin Havaflex T.A.-117 A Trowelable Ablative Material Rev. 6/21/06
This information set forth herein is furnished free of charge and is based on technical data which AMETEK believes to be reliable. It is intended for use by persons having technical skill and at their own discretion and risk. Since conditions of use are outside our control, we make no warranties, express or implied, and assume no liability in connection with your use of this information. Nothing herein is to be taken as a license to operate under or a recommendation to infringe any patents.

455 Corporate Blvd., Newark, DE 19702 U.S.A. • Tel: 800-441-7777 or 302-456-4431 • Fax: 302-456-4444
www.ametekhaveg.com • E-Mail: info.haveg@ametek.com

Data sheet for SCM3404//America/eng,revision no 1

GE Silicones

SCM3402/SCM3404/SCM3405/SCM3408

SCM3402, SCM3404, SCM3405, SCM3408
Silicone Coating Material for Roof Application

Product Description SCM3402/SCM3404/SCM3405/SCM3408 are one-part silicone materials specifically designed for use in roofing applications as the weatherseal coating in silicone coated, spray applied urethane foam roofing systems. SCM3402/SCM3404/SCM3405/SCM3408 are specified for use with roofing granules in the top coat when used in GE's roofing system and applied by a GE authorized applicator.

Key Performance Properties SCM3402/SCM3404/SCM3405/SCM3408 are one component materials requiring no mixing and can be applied by spray, roller, or brush. Fast cure rate, 15 to 30 minute skinover time provides adequate time for granule application. 1-2 hour tack free time assures adequate cure at the end of the application day.

The low spraying viscosity of SCM3402/SCM3404/SCM3405/SCM3408 assures easy spraying with airless spray systems using a minimum 25:1 ratio air motor driven pump. Cracks and grooves fill easily. High static viscosity assures material stays where it is sprayed. It won't flow off high spots or vertical areas on parapets.

SCM3402/SCM3404/SCM3405/SCM3408 are VOC compliant. 200g/liter mineral spirits content assures compliance in most areas of California, greater New York City, New Jersey (verify compliance in location of use).

Datasheet for SCM3404/America/eng, revision no 1

Typical Product Data

Property	Value	Test Method
Density, lbs/gallon	10.45	WPSTM P-14
Color SCM3404	Medium Gray	Visual to Standard
SCM3408	Dark Gray	WPSTM C-19
SCM3405	Tan	
SCM3402	White	
Solids Content, % by volume	65	
% by weight	80	
Tack Free Time, hours	2	WPSTM E-86
Skinover Time, minutes	30	WPSTM C-560
Viscosity, centipoise	9000	WPSTM C-560
Tensile Strength, psi	200	ASTM D-412
Elongation, %	400	ASTM D-412
Peel Strength, lbs/inch	27	WPSTM C-628
Hardness, Shore A	32	ASTM D-2240

Specifications

Typical product data values should not be used as specifications. Assistance and specifications are available by contacting GE Silicones at 800/255-8886.

® SilPruf is a Registered Trademark of General Electric Company.
™ UltraPruf is a Trademark of General Electric Company.
® TEFLON is a Registered Trademark of DuPont.

Instructions for Use

SCM3402/SCM3404/SCM3405/SCM3408 may be applied by spraying, rolling or brushing to clean, dry, structurally sound surfaces. For best results ambient temperature should be between 50 and 80 degrees F. Lower temperatures will lengthen the skinover, tack free, and ultimate cure time. Higher temperatures will shorten the cure time and working time for proper application of granules before the material has skinned. Sudden temperature declines may also result in dew formation on surfaces which can prevent adhesion development. For best results, relative humidity should be above 20%. Lower humidity will slow the cure rate significantly.

Care should be taken to avoid overspray. All overspray should be cleaned up immediately and before it has cured by wiping alternately with mineral spirits and dry rags. Cured material may have to be scraped off surfaces with a razor blade or scrubbed off with steel wool and mineral spirits. To control overspray, avoid spraying in high winds (above 15 mph) and/or mask all surfaces which are not to be coated.

Cleanup of equipment containing uncured material may be accomplished by flushing with mineral spirits. SCM3402/SCM3404/SCM3405/SCM3408 should not be left in pumping equipment and hoses for prolonged periods of time

http://www.gesilicones.com/USADataSheets/223.html (2 of 4) [8/27/2003 3:11:59 PM] unless all hoses are TEFLON® lined, all piping connections are sealed with teflon and all pump seals are teflon. SCM3402/SCM3404/SCM3405/SCM3408 cures by reacting with moisture. Equipment without teflon lining and seals will transmit sufficient moisture vapor to gradually form cured material on hose walls and at unsealed connections. This may result in increasing operating pressures and material flow restriction.

SCM3402/SCM3404/SCM3405/SCM3408 are identical products except for color. When applying in two or more coats it is advantageous to use alternating colors to assure uniformity of coverage.

SURFACE PREPARATION

SCM3402/SCM3404/SCM3405/SCM3408 may be applied over properly cured, structurally sound, clean and dry polyurethane foam. Loose particles of foreign matter should be blown, brushed or vacuumed away. Surface must be smooth and flat enough to prevent puddling of water on the finished roof. Large holes should be patched with polyurethane foam, allowed to cure then ground off flush with adjacent surfaces. Smaller holes may be filled with SilPruf® or UltraPrufTM sealants before application of the roof coating. Before applying roof coating be sure surface is also free of moisture and that at least 2 hours of adequate temperature and humidity remain before the onset of nightfall or inclement weather.

SCM3402/SCM3404/SCM3405/SCM3408 may be applied directly over each other and over most other silicone roof coatings. When recoating older silicone coated roofs be sure the surface is clean, dry, structurally sound and that all holes and underlying damage has been properly repaired. The existing coating may have to be power washed with an appropriate surfactant solution then power rinsed to remove dirt. It is recommended that a test patch be cleaned and coated with SCM3404/SCM3408 to verify the effectiveness of the cleaning process and the adhesion to the surface prior to beginning the job.

SCM3402/SCM3404/SCM3405/SCM3408 may be sprayed as received. Additional dilution with solvent is neither necessary nor desirable. On horizontal surfaces these materials may be sprayed easily in thicknesses up to 15 or 20 mils or more without flowing or alligatoring.

Consult the GE Silicones Roofing System Suggested Specifications For New or Retrofit Roofing, document #4549, for additional application information.

Handling and Safety Material Safety Data Sheets are available upon request from GE Silicones. Similar information for solvents and other chemicals used with GE products should be obtained from your suppliers. When solvents are used, proper safety precautions must be observed.

Datasheet for SCM3404/America/eng, revision no 1

Storage and Warranty Period — The warranty period is 6 months from date of shipment from GE Silicones if stored in the original unopened container at 25°C (77°F).

Availability — Products may be ordered from GE Silicones, Waterford, NY 12188, the GE Silicones' sales office nearest you or where appropriate, an authorized GE Silicones' product distributor.

Government Requirement — Prior to considering use of a GE Silicones' product in fulfilling any government requirement, please contact the Government and Trade Compliance office at 413-448-4624.

LEGAL DISCLAIMER

THE MATERIALS, PRODUCTS AND SERVICES OF GE SILICONES, GE BAYER SILICONES, GE TOSHIBA SILICONES, THEIR SUBSIDIARIES OR AFFILIATES (THE "SUPPLIER"), ARE SOLD SUBJECT TO THE SUPPLIER'S STANDARD CONDITIONS OF SALE, WHICH ARE INCLUDED IN APPLICABLE SALES AGREEMENTS, PRINTED ON THE BACK OF ACKNOWLEDGMENTS AND INVOICES, OR AVAILABLE UPON REQUEST. ALTHOUGH THE INFORMATION, RECOMMENDATIONS OR ADVICE CONTAINED HEREIN IS GIVEN IN GOOD FAITH, SUPPLIER MAKES NO WARRANTY OR GUARANTEE, EXPRESS OR IMPLIED, (I) THAT THE RESULTS DESCRIBED HEREIN WILL BE OBTAINED UNDER END-USE CONDITIONS, OR (II) AS TO THE EFFECTIVENESS OR SAFETY OF ANY DESIGN INCORPORATING SUPPLIER'S MATERIALS, PRODUCTS, SERVICES, RECOMMENDATIONS OR ADVICE. NOTHING IN THIS OR ANY OTHER DOCUMENT SHALL ALTER, VARY, SUPERSEDE OR OPERATE AS A WAIVER OF ANY OF THE SUPPLIER'S STANDARD CONDITIONS OF SALE.

Each user bears the full responsibility for making its own determination as to the suitability of Supplier's materials, products, services, recommendations or advice for its own particular purpose. Each user must identify and perform tests and analyses sufficient to assure it that its finished parts will be safe and suitable for use under end-use conditions. Because actual use of products by the user is beyond the control of Supplier, such use is within the exclusive responsibility of the user, and Supplier cannot be held responsible for any loss incurred through incorrect or faulty use of the products. Further, no statement contained herein concerning a possible or suggested use of any material, product, service or design is intended or should be construed to grant any license under any patent or other intellectual property right of Supplier or any of its subsidiaries or affiliated companies, or as a recommendation for the use of such material, product, service or design in the infringement of any patent or other intellectual property right.

technicalinformation
Inmont Corporation

1192 'MARTYTE' ABLATIVE COMPOUND

Martyte is supplied in three parts:

    Component A: Dry powder
    Component B: A proprietary epoxy resin
    Component C: A proprietary polyamide amine resin

The Material Safety Data Sheets are provided separately for each of these components. Please refer to these MSDS's for the hazards and safe and proper handling of these products.

These components are supplied in pre-proportioned units for easy field use. The compound can be mixed by hand or with small mechanical machines. When freshly mixed the material resembles a dry, powdery plaster. It is applied by troweling in place. Martyte provides a working time, or pot life, of approximately three hours at 77° F, and will cure completely within 18 hours at this temperature.

Martyte is an extreme-temperature resistant coating which was specifically developed to insulate rocket gantry and missile launch frames against heat, blast erosion and related deterioration.

A potential hazard may exist in the application and use of this product. Dust created during a launch or during removal and reapplication may present a hazard to those in the vicinity who are not wearing respiratory protection. It is recommended that an approved dust respirator be used to protect the employee from airborne particles.

INMONT

Inmont warrants that the product is of merchantable quality and will meet such specifications as have been agreed upon in writing. THE FOREGOING WARRANTIES ARE IN LIEU OF ALL OTHER WARRANTIES, WRITTEN OR ORAL, EXPRESS OR IMPLIED. UNLESS INCLUDED IN THE FOREGOING WARRANTIES, NO WARRANTY OF FITNESS FOR A PARTICULAR PURPOSE IS MADE. INMONT SHALL NOT BE LIABLE UNDER THE FOREGOING WARRANTIES OR ON ANY OTHER GROUND FOR DAMAGES IN AN AGGREGATE AMOUNT GREATER THAN THE PRICE PAID FOR THE PRODUCT.

PRESSTITE PRODUCTS
Inmont Corporation
1216 Central Industrial Drive
P.O. Box 509
St. Louis, Missouri 63169
Tel 314 577 1100

This page intentionally left blank.

www.ingramcontent.com/pod-product-compliance
Lightning Source LLC
Chambersburg PA
CBHW081735170526
45167CB00009B/3832